# 景观的延伸表现

## ——瓦莱里奥·莫拉比托的理念与方法

# The Extended Representation of the Landscape

## ——Ideas and Methodology of Valerio Morabito

项淑萍 Shuping Xiang
[意]弗朗切斯科·贝利杰兰特 Francesco Belligerante

中国建筑工业出版社

# 序 Foreword

瓦莱里奥·莫拉比托的画是美丽的心灵景观。

更确切地说，它们是记忆地图，是经过观察及切身体验后又特意离开现场而创作的。他的画是这样的地图，记录着想法如何塑造场所以及场所如何塑造想法。虽然这些画描绘的通常是城市和景观，但主要题材却是头脑中最神秘的涡流——记忆。无论作为艺术创作还是作为景观设计的方法，这都是极其重要的，因为它所承载的记忆和物质空间都是可以重写的。通过“画记忆”的方法，莫拉比托用心灵走进了场所深处。

他的画表现了碎片性的记忆本质，画也由此而生，比如扭曲的时空、触觉的回响、模糊的边界以及个性的倾向。在他的绘画中，莫拉比托把控着世界之主观性与客观性的平衡；而不会过于艺术而太主观（超现实主义），抑或者是过于设计而太客观（自然主义）。他在现实和表现之间的循环反馈中把控着，就像盘旋于科学家们所说的“奇异吸子”之间。

但这并不是一个本质主义的项目：他并不主张寻找自然、场所或是思想的原本真相。这些是发自无穷好奇心的愉快记录，知晓世界最终是未知的。然而，当他的双手与主观想法反复地交织在一起时，一些特定的图案形成了，如果你认真去看这些记忆地图，就会发现一些线索，它们导向新的景观形式。

理查德·韦勒
马丁和马吉·梅尔森都市主义主席
宾夕法尼亚大学景观系教授及系主任

Valerio Morabito's drawings are beautiful mindscapes.

More precisely, they are memory maps, intentionally created off-site and after the fact of observation and bodily experience has taken place. His drawings are maps of how the mind shapes place and how place shapes mind. While the drawings often depict cities and landscapes their primary subject is the mind's most mysterious vortex, memory. This is significant as an art practice and as a landscape architectural method because both memory and the physical places it records are palimpsests. Morabito uses his method of drawing-from-memory to pull himself deeper into a place and with it his own psyche.

His drawings are formed by and represent memory's fragmented nature, its warped space and time, its haptic reverberations, its blurred edges and its autistic tendencies. In his drawings Morabito holds the world in balance between the subject and the object; whereas so much art gives too much to the former (surrealism) and so much design too much to the latter (naturalism). He holds the world inside a feedback loop between reality and its representation, spiraling in on what scientists call 'strange attractors'.

But his is not an essentialist project: he is not claiming to find the truth of nature, place or mind. These are the pleasurable recordings of an iterative curiosity that knows the world is ultimately unknowable. And yet, as his hand weaves repeatedly through his subjects, certain patterns do emerge and if you look at these mnemonic maps carefully you will find clues that lead to new forms of landscape architecture.

Richard Weller
The Martin and Margy Meyerson Chair of Urbanism
Professor and Chair of Landscape Architecture at the University of Pennsylvania

# 卡尔维诺的遗产

1985 年，在卡尔维诺生命的最后时刻，他写下了新千年的五个备忘录，原本是为出席哈佛大学查尔斯·艾略特·诺顿讲座而准备的。但他没来得及写第六个备忘录就去世了，也未曾在美国有过任何讲座。后来，一些编辑收集了他的备忘录并出版成了书，意大利文版的《五篇美国讲稿》，或英文版的《未来千年文学备忘录》。

当我读到这些备忘录时，我异常惊讶。对我而言它透露出一个很不一样的卡尔维诺。我读了很多卡尔维诺的书，就像被诱拐到了一个独特的虚幻世界。我沉醉于令人难以置信的故事里，他居然能用如此简单又完美的意大利语描绘那些故事：他通过赋予灵魂将一座座新城创造得出神入化，他描述一处处风景而能将空间规模与感知延引至最深处的秘密，他还塑造诸如帕洛马尔先生那样神奇的人物角色。在他笔下，他总能用特有的方法将现实的平淡生活描绘成另一种现实；通过生活中小小的细节来揭示宇宙的奥秘，亦或通过宇宙中小小的物体来透析我们的生活。卡尔维诺写了许多帕洛马尔先生的故事，作为提高他意大利语写作技巧的练习。我通常会以大幅绘画的形式画画或做些大速写，以提高我的绘画技巧，也作为思考空间的一种方式，这便是我从卡尔维诺那里学到的很重要的一课。

读了新千年的六个备忘录之后（我们正在经历的千年），我惊讶于卡尔维诺强大而难以置信的文学技巧解析。他构造句子语法的独有方式建立在与其他一些作者的关联性之上。我之所以惊讶是因为这些备忘录是对于规则、构件、从句和创作要素的解释，而利用这些法宝，卡尔维诺创造了他自由、简单及准确的故事。轻逸、速度、准确、鲜明和多样是他解释他文学观点所用到的词汇，并是他建议我们在这个千年里使用的关键词。

在其中一个备忘录里，他谈到我们所居住的世界中泛滥的图像，他说与之抗衡的唯一武器就是“文学想法”。那段关于“想法”的特定描述深深印在我的脑海中：我们可以用它来界定文学、音乐、建筑、景观或艺术的独到想法，我们必须以自己的方式来创作。同时他在书中通过许多图像描述了大量城市和景观空间；所以他也没有回避它们，但他总是将其与自己的“文学想法”结合在一起。

从那时我开始思考，我们创造新的建筑或景观空间的方式可以分成三个阶段的想法：第一个是景观（或音乐、建筑、文学等）的“想法”；第二个与偶然的想法、故事、插曲或需改变的场地相关；第三个关系到我们讲故事时构造语法的能力。

第一个想法与我们的知识相关，我们怎样认识周围的世界，我们怎样在不同的文化领域构建联系，以及我们怎样使用一种方法去认识。每个人都有不同的方式实现这第一步，有些人非常精确，并能把所有的信息组织成一个完美的序列。但我更喜欢以跳跃的方式去认识，以故事、人物和场景作为空间中的点，由于我糟糕的记性，我需要将这些点与我的想象和创造结合起来，这是我需要解决的问题。

第二步是关于我们想表达什么，或者怎样才能创作图像；我们预想改变的场地以及预想形成的空间总是与景观（或音乐、建筑、文学等）的“想法”相关。改变场地及其特征、形态、个性、生态和环境是可能的，但是，我们必须用自己的“想法”为武器去反对制作一些平庸的图像。

第三步涉及我们处理空间语法的能力，那种我们将元素置入空间并创造连接性与一贯性的特殊方式。为了提高这种构成空间的特殊方式，我通常做一些画作练习，就像卡尔维诺收集帕洛马尔先生的故事那样。

所以我的画是关于将来的记忆。我从来不在景观现场画画：包括城市、自然或工业。通常都是在之后画的，没有实景在我面前；经过一段时间，记忆便开始以直觉的形式呈现景观未来的图景。若干线条之后，通过记忆进行的景观表现，变成了一种对不准确的视觉认知的处理。通过增加那些近乎现实却并不存在的元素，塑造一个不同的现实，画作也形成了自有的尺度关系。

在我看来，表现关乎于场地可以如何变得不同，它能够怎样改变。场地在草图中是活的，避开了现在，而通过记忆的加工进入到它的未来。草图删除了现在并决定了它不存在。关于现在的想法已纯然不存在了，它消失在记忆和它的未来中。

我的画并没有预设的技术，但会依据不同的场地自然而然的形成，之后，它们又自行建立准确的规则：以能与数码相机相媲美的速度将草图模拟和数字化，控制照片并利用它们创造新的关系。当场地变了，技术也随之改变并衍生出自己的规则。我尝试着避免画的重复性，利用错误的不确定性往往能带来最合理的精确。

后来，有些草图变成了绘画，但却不是一般意义上的绘画，它们更像是画布上的大草图——用了油料、丙烯或墨色，看上去像绘画。它们体现了不同尺度上表现材料和形状之技法的对比。

很多时候，这些画跟我景观项目中的设计方法无关，仅仅是我想训练编故事的技巧而留下的一些练习和小故事。这是一种诗意的工作方法，不论它们被用与否，一切都能够发挥作用：不管怎样，它们都是重要的。在这一过程中，草图有待于被利用到一个新的空间项目中，它们是场所精神，与目标场地的场所精神相违背的第二场所精神，能够解决场地形态与诗意、几何与幻想方面的问题。

瓦莱里奥·莫拉比托

美国宾夕法尼亚大学兼职教授

意大利地中海雷焦卡拉布里亚大学教授

# The Heritage of Calvino

Calvino, at the end of his life, in 1985, wrote five memos for the next Millennium, that were meant to be presented at the Harvard University for the Charles Eliot Norton Lectures. He passed away before writing the sixth one, and he had never had any lectures in the USA. Later some editors collected the memos in a book, "Cinque Lezioni Americane" in Italian, or "Six Memos for the next Millennium" in English.

When I read the collection of the memos, I was incredibly surprised; it was as revealing a different Calvino for me; I had read many of Calvino's books and I was kidnapped in a special unrealistic world. I was lost inside incredible stories, that he was able to conceive writing and picturing them with a simple and perfect Italian: the invention of fantastic new cities through the invention of their soul, the description of landscapes in which the scale and the perception of space were extended until their deepest secret, and the invention of incredible characters such as Mr. Palomar. Through him, he was able to describe the reality of the common life using a method to pass from one reality to another; by means of small details to reveal the secret of the universe or, starting from the inner universe, he was able to pick up one small object which reveals our common lives. Calvino wrote the stories of Mr. Palomar as a collection of many exercises that he did to improve his Italian technical skills. I usually draw or make big sketches, in form of big paintings, to improve my technical skills, a way to think through spaces, and it was an important lesson I had from him.

When I read the six memos for the next millennium (the millennium in which we are living today), I was surprised by the strong and incredible technical explanation about literature; the connection and the links to several authors were a basis that he used to build his own way of composing the grammar of every sentence of his. I was astonished because the memos were an explanation of rules, devices, clauses and compositions to be used to create a free, simple and precise invention of stories. Lightness, Quickness, Exactitude, Visibility and Multiplicity are the words he used to explain his vision about literature, and to suggest us which key words we could use in this millennium.

In one of them he spoke about images that crowd the world in which we live, and the only weapon he has to oppose them, he said, is the "idea of literature". That particular description of the meaning of "idea", was a strong passage in my mind: we could use it to identify ideas of literature, music, architecture, landscape or art, that we must use to compose our own way. At the same time he uses many images in his books, describing a lot of spaces of cities and landscapes; so he does not avoid them but he always connects them with his own "Idea of Literature".

From that moment I have thought that in our way of inventing new spaces of architecture or landscape, we could have three steps of ideas: the first one is the "idea" of the landscape (or music, architecture, literature and so on), the second one is related to contingent ideas, stories, episodes or sites which we have to modify, and the third one is linked to our capability to deal with the grammar of our composition in telling stories.

The first idea is related to our knowledge, how we know the world around us, how we create links with different fields of our culture, and how we use a methodology to be able to know. Each of us has different ways to approach to this first step, someone is very precise and able to put all the information in a perfect sequence, but I prefer to know by jumping from one information to another, using stories, characters and situations as points in the

space that I have to link with imagination and inventions because of my bad memory, with which I have to deal.
The second step is related to what we want to speak about or how it is possible to produce images; the site we want to change, the space we want to form are always related to the "idea" of the landscape (or music, architecture, literature and so on). It is possible to change the site and its features, the morphology and the identity, the ecology and the environment, but to oppose to the images that we must produce, we have the weapon of our own "idea".
The third step is our capability to deal with the grammar of the space, the particular way we put together the elements into the space creating links and consequentiality. In order to improve this particular way of composing the space, I usually draw doing exercises, like Calvino did when he collected the stories of Mr. Palomar.
So my drawings are a memory of future. I have never drawn in front of a real landscape: urban, natural or industrial. I always draw the landscape later, without having it in front of me; as I wait, the memory represents the future trying to build a form of intuition of it. After some lines, the drawings, which represent a landscape through memory, become a non precise visual perception you have to deal with. A different reality, shaped by a process which adds not existing elements as they are real and the scales of the drawings become self references.
In my opinion the re-presentation is related with how a site could be different, how it could change. The site is alive in the sketch drawings and, processing the memory into its future, it avoids the present. The sketch drawings delete the present and decide that it does not exist. The idea of the present simply does not exist anymore, it disappears into the memory and its future.
In this way each sketch drawing analyzes the reality and produces its design: in my mind the memory is made by spots, it is fragmented. For this reason my memory produces extensive blank spaces to be filled, voids in which the future stretches out inside them and fills them like water.
My drawings do not have a predetermined technique but they choose one in relation to the different sites they pretend to represent and then, after deciding the technique, they build precise rules by themselves : analog or digital the sketch drawings would like to compete with the speed of the digital camera, controlling its photos to use them to create new relationships. When I change the site, the technique changes and develops its rules. My drawing tries to avoid the repetition and uses the precarious of the mistake: the best possible precision.
Afterwards a few sketches become paintings, although not in the common sense of the term, but they become more like big sketches made on canvas - with oil, acrylic or ink colors - which look like paintings; they are expressions of the contrast between the different scales of the technique, in terms of materials and shapes.
Many times these drawings are not related to the design approach in my landscape projects, but they are exercises, short stories only written because I want to train my skills to write stories. This is a kind of poetic methodology of work, in which everything could help either they are used or not: anyway they are all important. With this process the sketches wait to be used in a new space as projects, they are the Genius Loci, the second Genius Loci opposed to the Genius Loci of the site, able to deal with it in terms of morphology and poetry, geometry and illusion.

Valerio Morabito
Adjunct Professor at University of Pennsylvania, USA
Professor at Università Mediterranea di Reggio Calabria, Italy

# 序 Foreword



# 起源 Origin

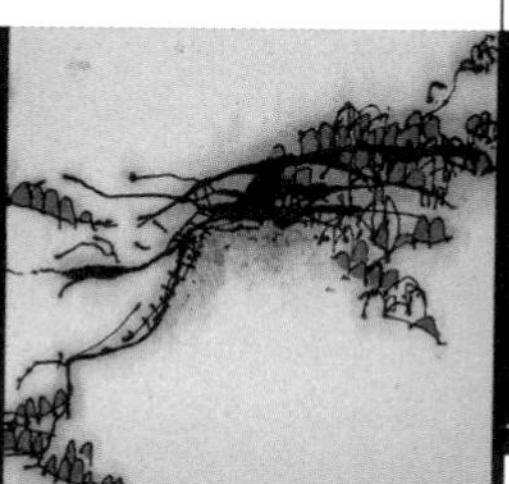
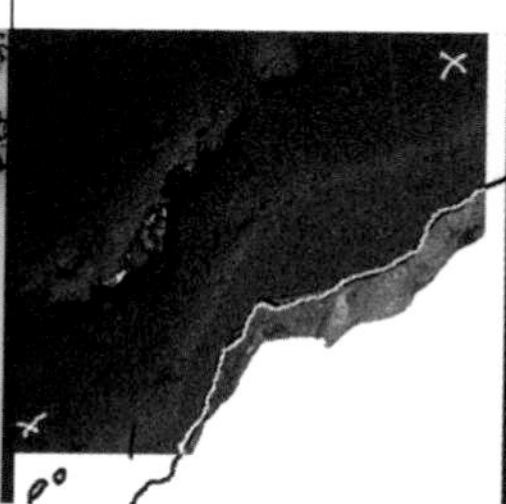

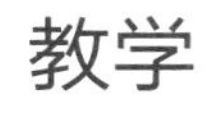

# 目录 Contents

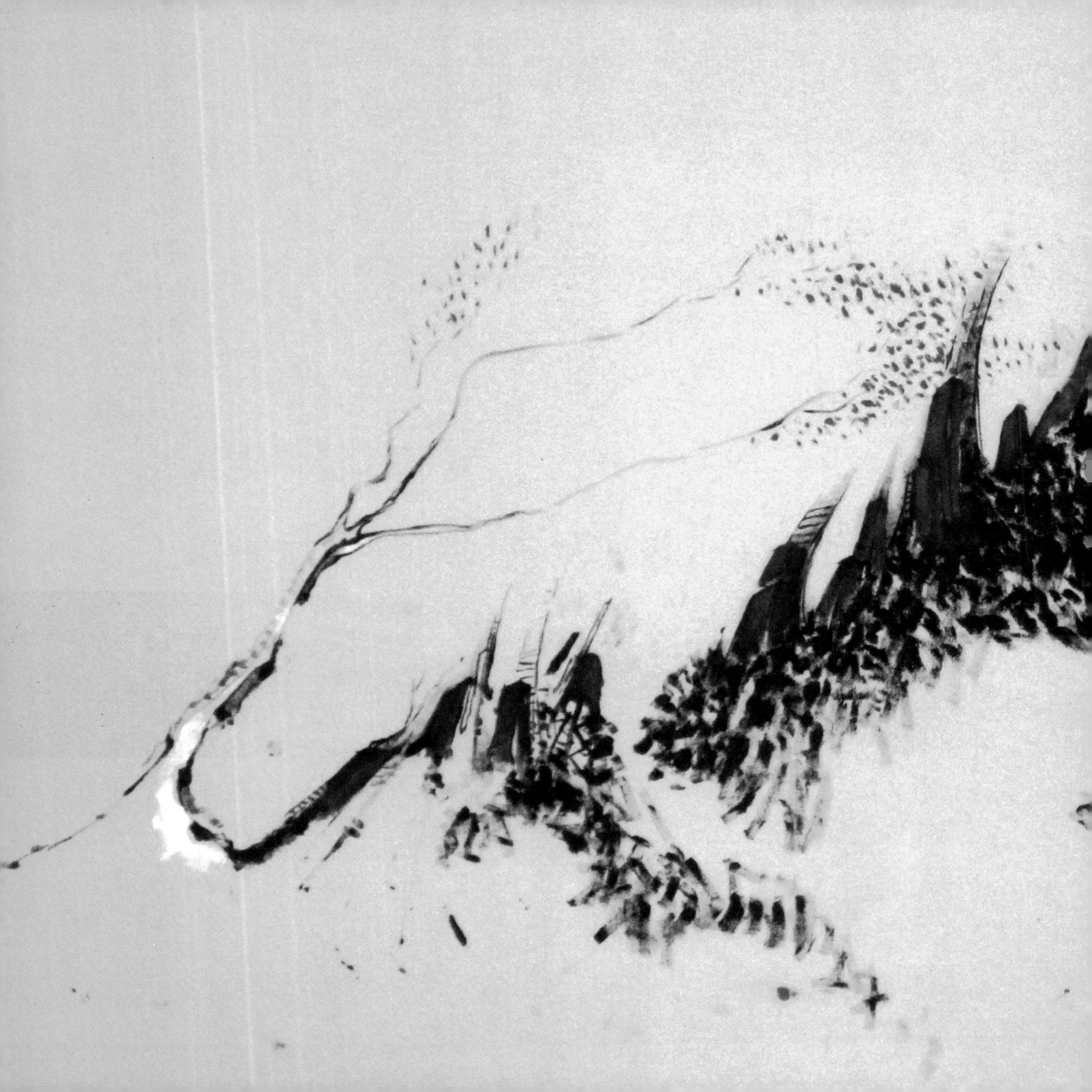

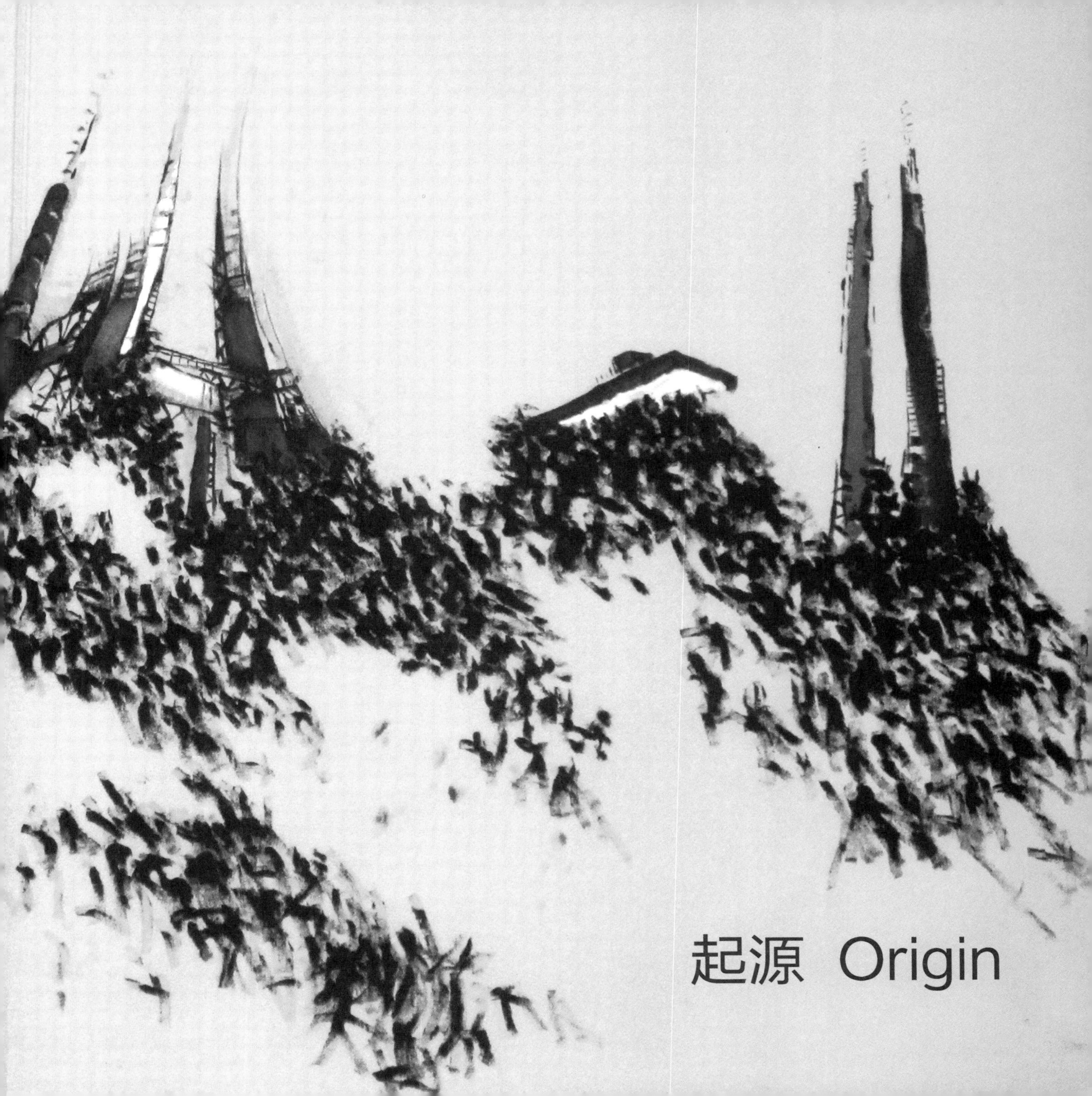

# 起源 Origin

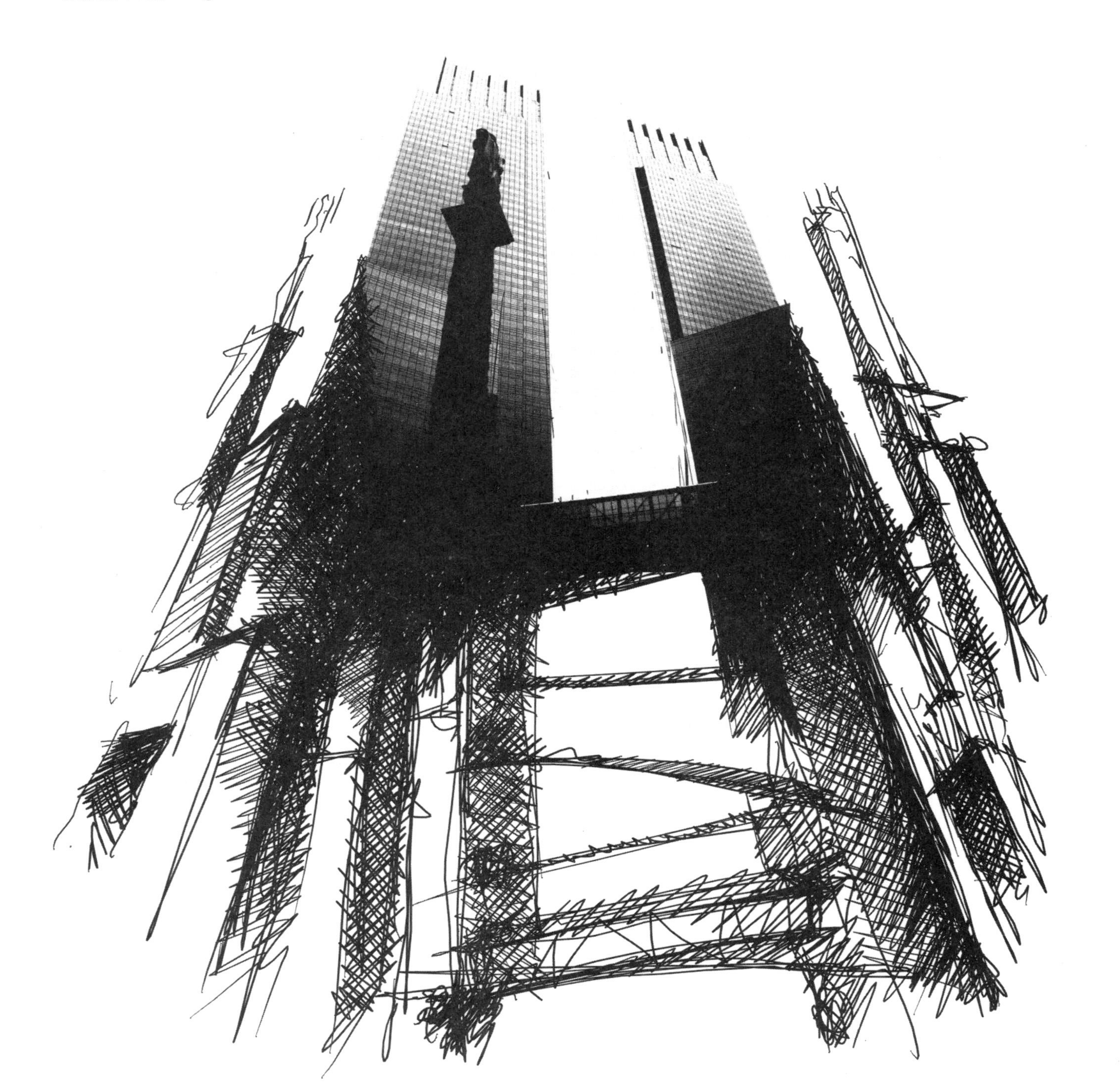

# 延伸表现方法
# Method of the Extended Representation

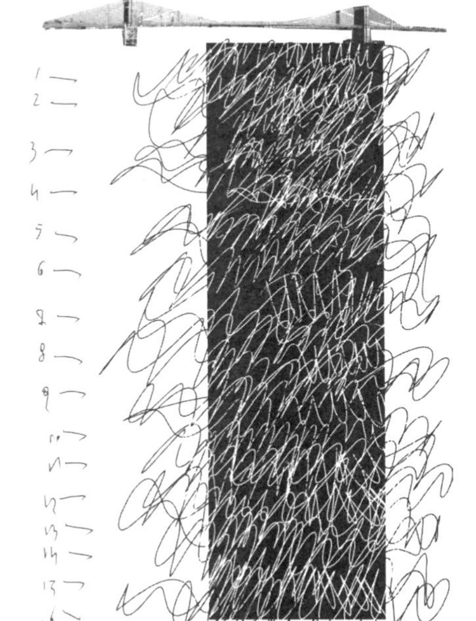

## 一个故事的启发
## Inspiration from a Story

不论是什么，要掌握一种新方法需追根溯源，理解令其诞生的动力。“延伸表现”的开创者瓦莱里奥·莫拉比托教授非常热衷于文学。在众多的作家中，20世纪最著名的意大利作家之一伊塔洛·卡尔维诺对他的影响最突出，并一直陪伴着瓦莱里奥的专业成长。一个出自卡尔维诺《圣约翰的街道》的小故事“源自不透明性”，开启了瓦莱里奥对空间和表现的全新想象。

To master a new methodology, whichever it is, starts from understanding the motivations that lead to its birth. Professor Valerio Morabito, who invented the “Extended Representation”, has a great passion for literature. Among numerous writers, Italo Calvino, one of the most important Italian writers of the twentieth century, gives Valerio the greatest influence and accompanies his professional growth. Inspired by the writer's short story entitled "Dall'opaco", from the book *Le strade di S.Giovanni*, Valerio was driven into a new and innovative vision of space and representation.

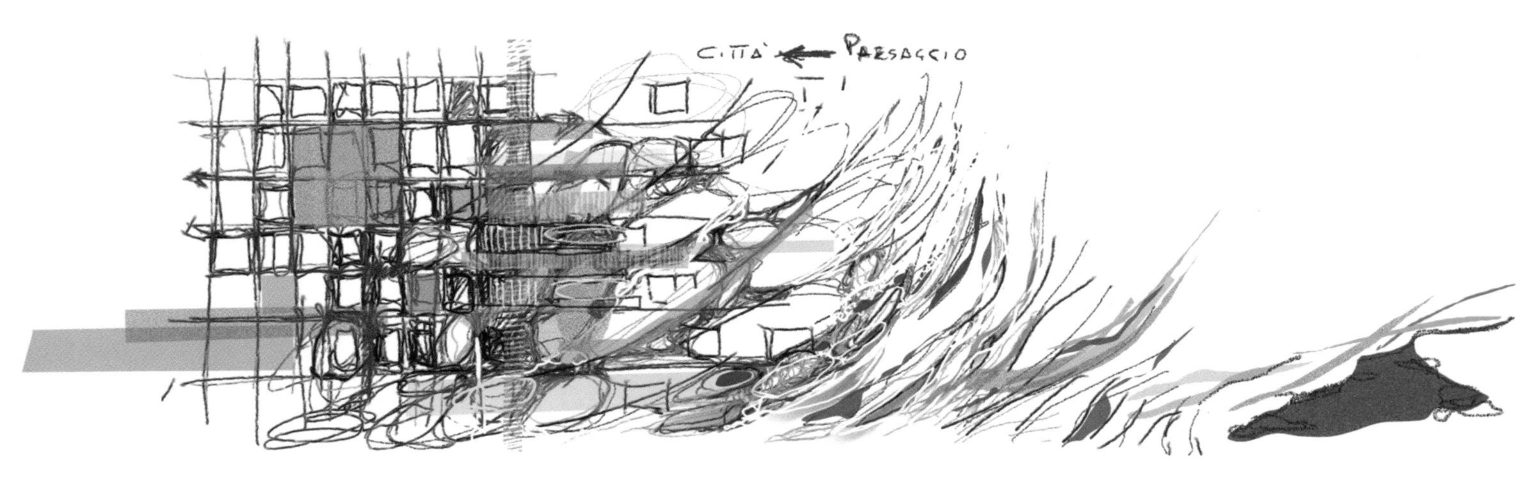


## 源自不透明性

“……如果我被问到空间到底有多少维度，如果有人问如何在缺乏这些知识的情况下尝试建立一套通用的规则来认知空间，其中第一条就是每个人都被三个无限维度穿越的概念。一个从我们的胸膛进入并穿过后背；另一个从我们的一个肩膀横穿另一个肩膀；第三个钻过我们的头颅从脚底出去。只有一个历经重重质疑和拒绝后才被接受的观点，人们才能习以为常地去使用它。若要把我的答案建立在我如何用之观察四周的基础上，需建立在那三个维度上。当你置身中央时，它们真的变成了六个，前面后面上面下面左面右面。当我说着这些话并将我的脸转向大海而背向着山时，我关注着它们。第一点想说的就是我前面的维度实际上并不存在，就像我面前下方突然变成了空无，然后变成了大海，然后变成了地平线，然后变成了天空。所以我们也可以说我面前的维度与我上方的维度是一致的，与你直立时从你的头颅当中伸出来的那个维度一致，并且它瞬间消失于新的顶点，将继续移至后面的维度，因为它碰到了一堵墙、一块石头、一个陡坡或者一丛灌木。如我所说的，我背对着山，所以我也可以说这个维度不存在或者融入了在我下方的维度，这个维度本应当从你的脚底延伸出来，但事实上根本没有，因为在你的脚底和铺装之间没有足够的空间容许它伸出来。然后是从左边和右边伸出来的维度，就我而言这与东西向的维度或多或少是相同的，而且这个维度能够向两个方向同时延伸。因为世界以高低错落的环境呈现，所以每个高度都可以描绘一根虚构的水平线，切入世界的斜面，就像被画在高程地图上的那些线，它们还被赋予了可爱的名字——等高线……”

——伊塔洛·卡尔维诺

# *Dall'opaco*

*"……if I had been asked then how many dimensions there were to space, if someone had asked that self of mine that still lacks knowledge of those things that are learnt to create a code of shared conventions, the first of which is the conception that each of us is crossed by three infinite dimensions, by one that enters through our chest and then comes out through our back, by another that crosses us from one shoulder to another, and by a third that drills through our skull and comes out of our feet, an idea that one accepts after many doubts and refusals, only to then act as if one had always known it, if I had to base my answer on what I had really learnt looking around me, on the three dimensions which, when stuck there in the middle of it all, really mutate into six, in front behind above below to the right and the left, looking at them as I was saying with my face turned towards the sea and my back to the mountains, the first thing to say would be that the dimension in front of me does not exist,as what is down there before me suddenly becomes nothing and then becomes the sea, which then becomes the horizon, which then becomes the sky, so we could also say that the dimension in front of me coincides with the one that is above me, with the dimension that comes out of the middle of all your skulls when you are standing up straight and that is suddenly lost in the new zenith, and would then move on the dimension that is behind because it comes up against a wall, a rock, a steep slope or a shrub, as I say always with my back to the mountains, in other words at midnight, so I could also say that this dimension doesn’ t exist or blends into the underground dimension below me, with the line that should come out of the soles of your feet but in fact doesn’ t come out at all because it doesn’ t have enough natural space to come out between the soles of your shoes and the pavement, and then there is the dimension that stretches out to the left and to the right, which for me is more or less the same as to the east and to the west, and this one can continue in both directions because the world proceeds with its jagged environment so that at every level it is possible to trace an imaginary horizontal line which cuts into the oblique slope of the world, like those that are drawn on top of altimetric maps and which go by the lovely name of isohypses……"*

——Italo Calvino

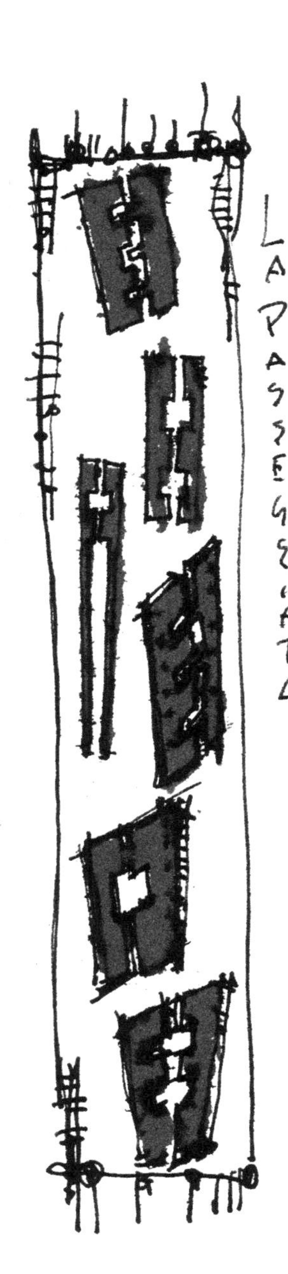

在这个故事中，一个众人皆知的由抽象 XYZ 轴构成的坐标系，被创造性地用来阐释现实世界中的空间。但是，不同于笛卡尔和欧几里得模型，卡尔维诺引领我们凭借人类的直觉进入一个富于想象的世界，这远不同于数学家们通常所勾勒的理性与固定的世界。利用我们自己的身体作为一个活的参照系，卡尔维诺提供给我们一种现实而新颖的阅读空间的方式，根据个人对空间的体验和解读，获取一个独特的、动态的、梦幻的图景。由于景观空间的模糊性和抽象性，对于景观的表现不适合完全利用笛卡尔和欧几里得的理性方式，景观表现势必要突破其限制而得到延伸。瓦莱里奥将这一过程的实现归纳为三个方面进行阐述。

In this story, a familiar Coordinate System with the abstract XYZ axis, is creatively employed to interpret the space in the real world. Differently from the Cartesian and Euclidean models, however, Calvino guided us into an imaginative world with human intuition, which is far beyond the rational and fixed world that the Mathematicians usually figure out. With a live dimensional-system formed by our own body as a reference system, Calvino provided us a realistic and yet innovative way to read the space, in order to get a unique, dynamic and fantasy picture according to personal experience and interpretation of the space. For the vagueness and abstractness of the landscape, it is not suitable to represent it in the rational way of the Cartesian and Euclidean school traditions, rather, the representation of the landscape is forced to break its limitation into extension. The implementation of this process is explained by Professor Valerio in three well-defined aspects.

# "延伸"

延伸也可以说是“延长”，即放大一个想法，扩展一个概念或简单的关联于某一事物。延伸就是这样一个举动，它有助于将空间关联到周边环境，或将我们周围的空间关联到新的想法、概念或简单事物。

景观因此变成了现实空间的延伸，表现因此成为一幅延伸的图景，通过描绘现状，在现实的景观本身之上融入它的标记，以成为今后景观的一部分。

以这一方式转变而来的图像属于景观的一种文化选择：作为一个抽象的概念，图像中的想法弱化了单个图像的具体含义，而使之具有更为复杂和总体层面上的意义，因此，没有必要对图像进行个体的评判；相反，重要的是创造它的隐性的复杂网络关系。

景观的延伸表现图不再局限于对现实数据的复制，或在景观项目中机械地表现一个想法，而是在景观创造的实践中植入其重要的作用。它们是包括了场地分析的表现图，而同时也蕴含了改变、增加体验，并进行材料归类以创建一个景观延伸表现的档案库。

# "Extension"

Extension is that word that "lengthens", enlarges an idea, stretch a concept or simply connects an object. Extension is the act that helps to relate space to its surroundings, or to relate the space around us to new ideas, concepts or simply objects.

The landscape thus mutates into the extension of real space and represents the condition by which representation as an extended image makes its mark on the reality of the landscape itself, in its aim to become part of a subsequent appearance.

The image transformed in this way belongs to the landscape as a cultural choice: the idea of the image as an abstract concept impeding a specific meaning for each image has a more complex and general meaning, in which the image has no need for individual justification, alluding instead to the complex network of relations which created it.

The images of the Extended Representation of the Landscape are no longer confined to the reproduction of real data or the representation of a landscape project as a real representation of its idea, but are contextulised in their vital function of being produced as exercises in landscape. They are images of a representation which contain analyses of the site, while at the same time containing its transformation, adding experiences and providing a stimulus for the cataloguing of materials to build a form of Extended Representation of the Landscape archive.

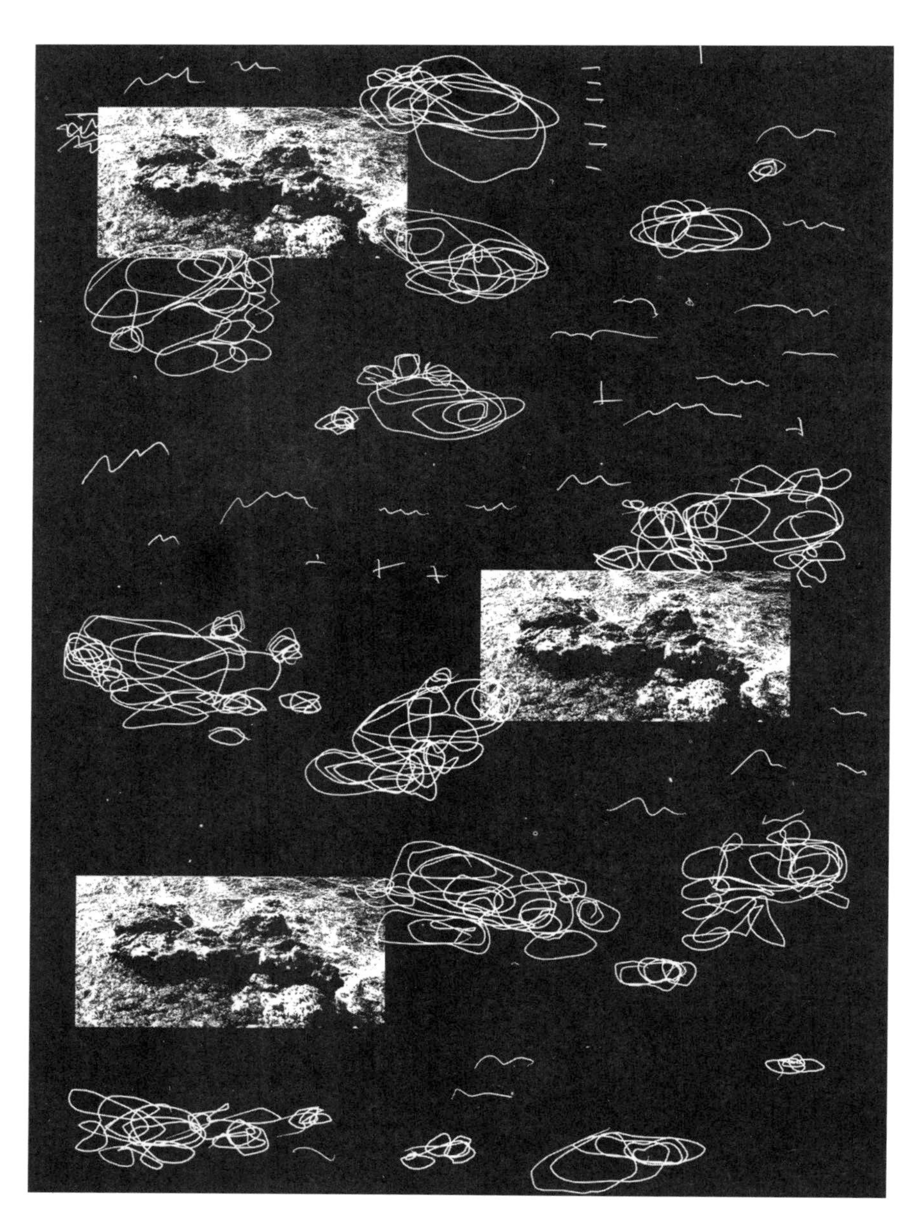

## "练习"

作为一种直觉而非理性认知的过程，介绍练习时不可疏漏徒手画这一有些脏且不完美的过程。

多年来，由于主观性的行为和对于绘画美丑意见的分歧，徒手画的草图已经被视为一种限制。

但如果草图能够延伸出去获取更多的信息，或它延伸我们已取得的信息，那么它不再局限于美丑之分。

在表现过程中它具有了新增的功能，用一种新的更灵活的研究方法，延伸每个要素的知识，对现实进行一种与众不同的表现。

草图因此失去了它的特性，且不再独自呈现于白纸前。而恰恰相反，由于外部数据的支撑，它所继承的几何图形消失在演变的过程中，并成为非欧几里得式延伸中复杂网络的主导因素。

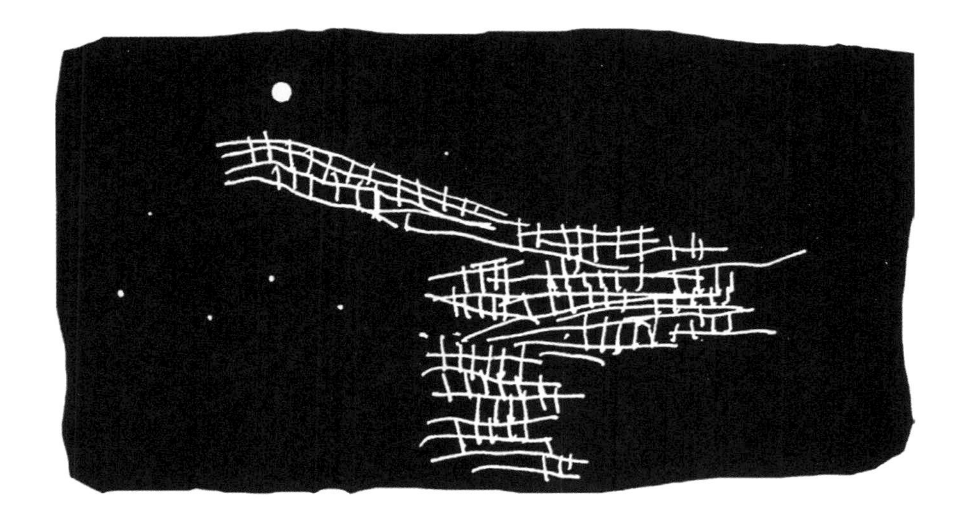

## "Exercise"

The introduction of exercise as an intuitive rather than rational cognitive process cannot fail to include drawing by hand, the somewhat dirty and imperfect process of the freehand drawing.
For years the freehand sketch has been seen as a limitation due to the subjectivity of movement and the division of opinion on which drawings are beautiful and which are not.
But if the sketch extends and reaches out for more information, or if the sketch extends the information acquired by ourselves, then it is no longer confined to the category of beautiful or ugly.
It has an added function in a representation process which extends its knowledge to each element for a different representation of reality, in a new, more flexible investigative method.
The sketch therefore loses its characteristic, and is not alone in front of the blank page, but quite the opposite. Supported by external data, it inherits geometries which disappear through the evolution of the process, and becomes the director of a complex network of non-Euclidian extensions.

# "记忆"

景观延伸表现的另一个基本的方面是记忆：不准确的、点式的、布满了漏洞……

我们的记忆存在漏洞是正常的，这些在现实的点与空间之间的无法则的空白，似乎出自赋予每一样事物以外观的未知理性法则，正如我们所说的，这些空白是景观的延伸表现最重要的延伸之处。

当我们处于想表现的对象面前时，我们习惯性地试图汇集尽可能多的信息去度量空间，然后期望将之量化，但如果在我们“看到”一个地方之后，先等一段时间再去表现它，那么记忆中的练习就不再仅仅与我们所看到的相关，而也与我们所知道的，所有不同的视觉体验相关，对我而言，最重要的是文学。

如果景观的延伸表现是通过这种方式完成的，那么它是有生命力的，前提是我们愿意注入一股连续的未经分类的信息流，它们互不相关、不准确、是支离破碎的，甚至是错误的。这些累积的东西，无法立即控制它们，随着时间演进便会找到准确的几何归属感，通过一系列的法则可能让任一信息自动重获归属，并通过我们控制那些空白空间的过程使之合理化：如此我们就能凭着记忆而非视觉进行表现了。

在基于量化城市记忆的传统景观分析中，融入这种方式，知觉的、个人的和直观的分析被结合在一起，依据“延伸”的法则能够获得与“客观”分析同样的科学价值 。

更重要的是它们成为归属过程中的重要基础，目前已经非常清晰的是无论哪种景观，空气或土地生态，凭其自身不能带来创新。

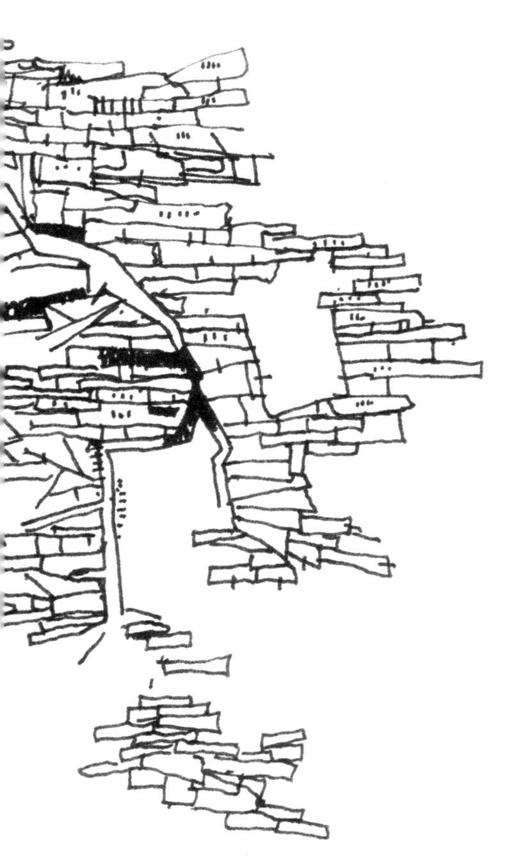

# "Memory"

Another fundamental aspect of an Extended Representation of the Landscape is memory: inaccurate, by points, filled with holes……

And it is normal that the holes in our memory, those non-coded blank spaces between real points and spaces, which seem to be invented by rational codes which give everything the appearance of having been made following unknown rules, these white space, as we were saying, are where the most important extensions of the Extended Representation of the Landscape are formed.

When we find our-self in front of an object that we wish to represent, we habitually try to bring together as much information as we can to measure the space and then quantify it in prospective terms, but if, after we have "seen" a place, we wait a while before representing it, then the exercise in memory is not only related to what we have seen, but also to what we know, to all those different visual experiences that, for me, come above all from literature.

If the drawing of the extended representation of the landscape is done in this way, it survives, as long as we are willing to feed it with a continuous flow of uncatalogued pieces of information which are unrelated to each other, imprecise, fragmented or even incorrect. This whole accumulated mass, uncontrollable in its immediacy, will with time find a precise geometry of belonging, a series of codes through which it is possible to automatically retrieve any piece of information and legitimise it by means of processes that each of us regulates within those blank spaces: we can then develop representation through memory and not through vision.

By acting in this way in traditional analyses of the landscape with a quantitative urbanistic memory, sensorial personal and intuitive analyses are joined together, gaining the same scientific value as "objective" analyses in terms of the "extension" of codes.

What is more, they become fundamental in a process of belonging in which it is now clear that knowledge of any type of landscape, of any air or land ecology, cannot in itself bring innovation.

# 瓦莱里奥的作品
## Valerio's Works

瓦莱里奥多年探索的绘画作品，直白地述说着他独特的方法，纸笔间透露出延伸表现的思想。“景观”、“城市”和“记忆”的三大分类关联于场地不同的特征，遵从源自事物与场地的逻辑图式，从中我们既能领会延伸表现的统一理念，又能察觉各类别的不同表现技巧。瓦莱里奥所有的作品和成果都包含着一个明确的过程，即“延伸”思想的设计，这一过程也展现了他是如何利用记忆进行绘画的，而这种记忆往往渗透着源自他知识背景里的新信息。

此外，从徒手画到油画，从数码图像到图片拼贴，对于多样表现技巧的尝试，使延伸表现方法一直处于发展演变的过程中。

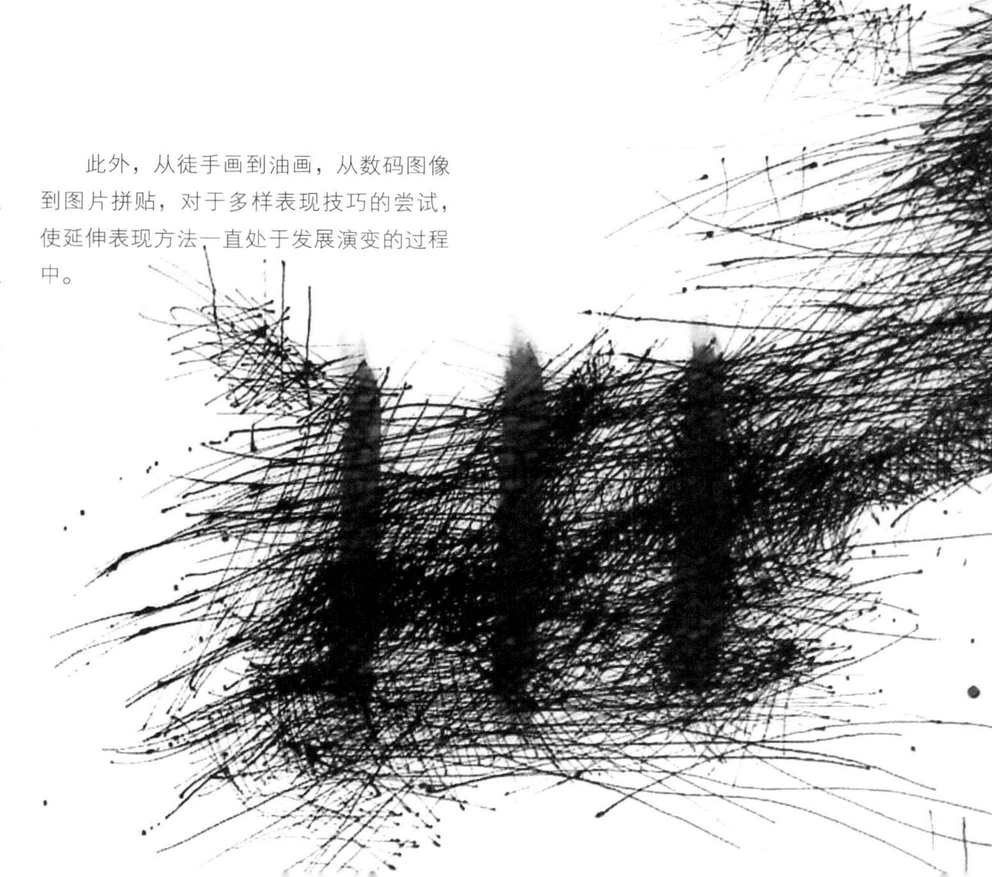

Valerio's works, with his exploration over time, explicitly restate his unique methodology, alluding the idea of extended representation through the strokes. "Landscape", "city" and "memory" are three categories characterized by different features that refer to the place, and they follow a logical pattern that stems from the subjects and places from which we are able to see the general idea of extended representation and the technical differences of each category. All of Valerio's works and accomplishmentsfollow a defined process of the design of the "extended" idea, which also shows how he works with his memory, that is usually mixed with the new information from his knowledge. Besides, from freehand sketches to oil painting, from digital images to photo collages, the use of various representative techniques allows the methodology to evolve over time.

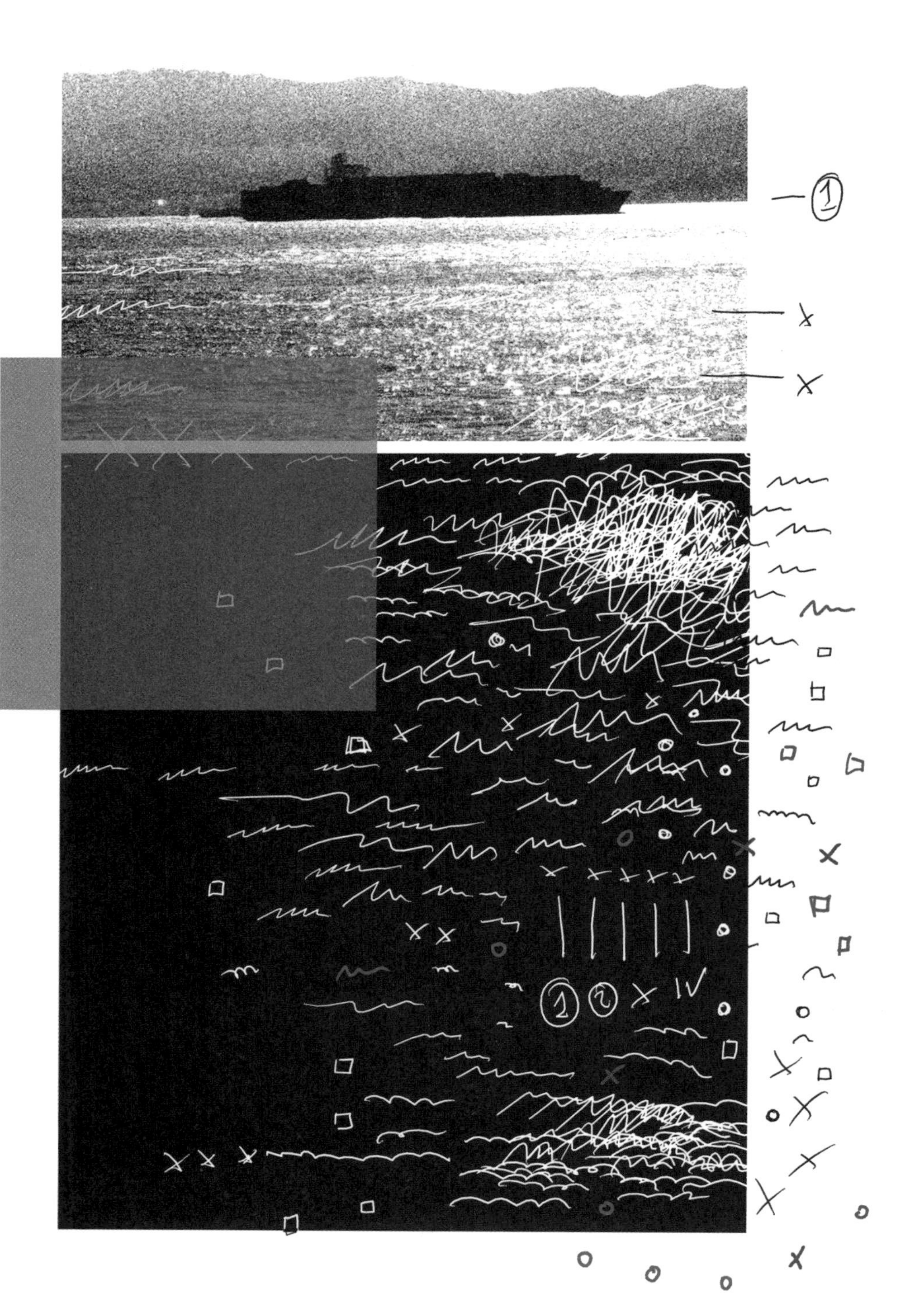

# 景观中的记忆
# Memory in the Landscape

通常在旅行之后，对于留下深刻记忆的自然景观，瓦莱里奥开始琢磨如何用最恰当的方式对它进行表现。灵感与创作往往需要这样一个过程：过滤繁杂信息达到最少，提取主要特征和本质要素，重组留在记忆中的符号和信息，用新的思想观念进行个性化表现……或许，这种过程不能就此简单地一概而论，但他总是有能力把复杂的景观空间描绘到极致的抽象、简明而富于诗意。通过瓦莱里奥的眼睛和记忆，“景观中的记忆”作品将带领我们穿行于他的自然景观世界，进行一个特殊的旅行，体验他对海与岛、山与农业、沙漠与地形等的个性描述。

After traveling through the natural landscape which leaves impressive memory, Valerio starts to think about the best way of representing it. More often than not, it takes such a process to capture the inspiration and to start the composition: filtering the mass information to get to the minimal essence, capturing the general characteristics and essential elements, reorganizing the signs and information remained in memory, and personalizing the representation with a new idea and vision…… Maybe the specific process is far more than just a simple summary, but Valerio is able to interpret the complicated landscape into an extremely abstract, simple, precise and poetic form. His work “Memory in Landscape” will guide us into Valerio's world of natural landscape, through his eyes and memory, guiding us into a special trip, experiencing his personal narration about the sea and the island, the mountain and the agriculture, the desert and the topography, etc.

加那利群岛，西班牙  Canarian Island, Spain

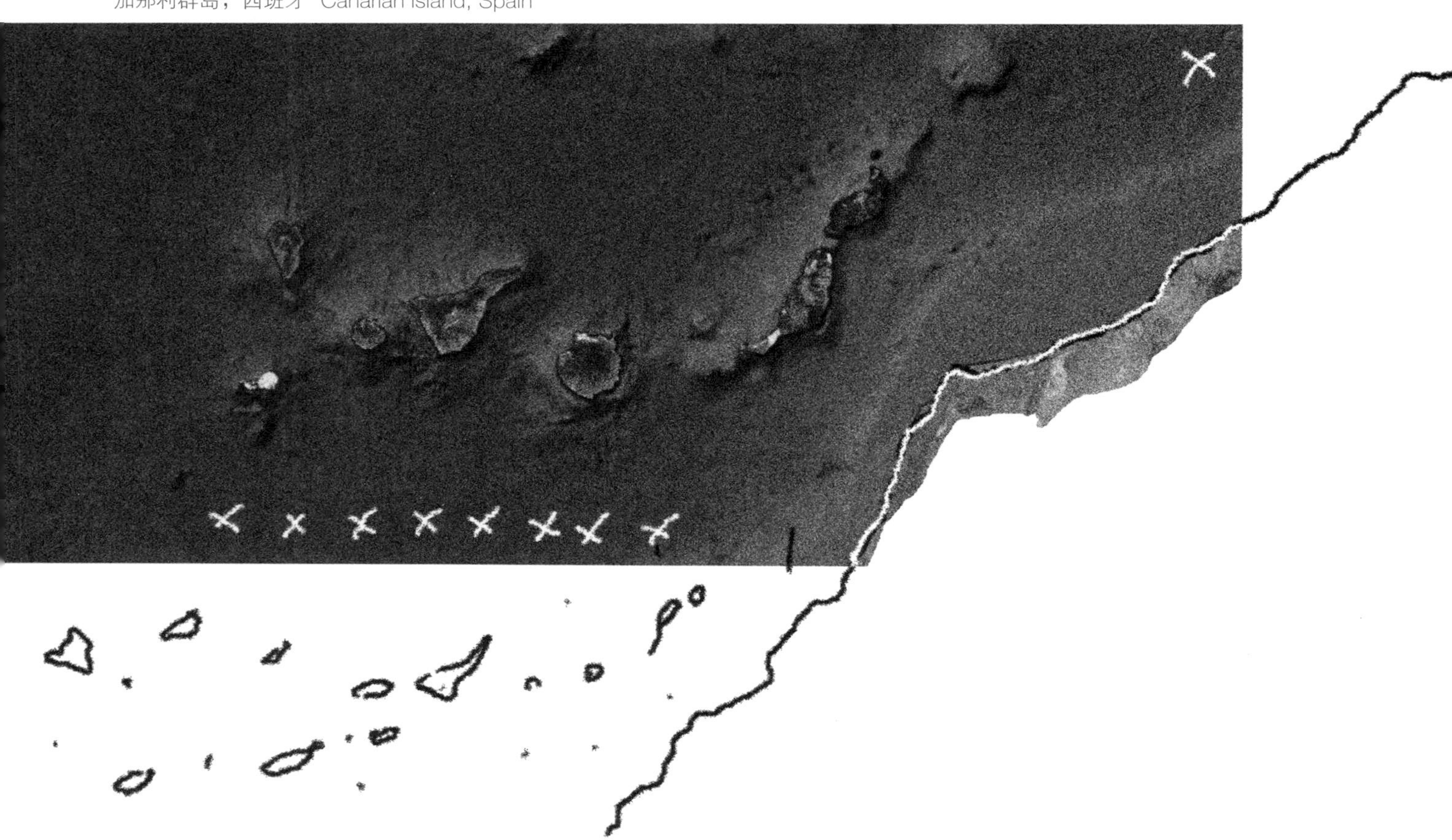

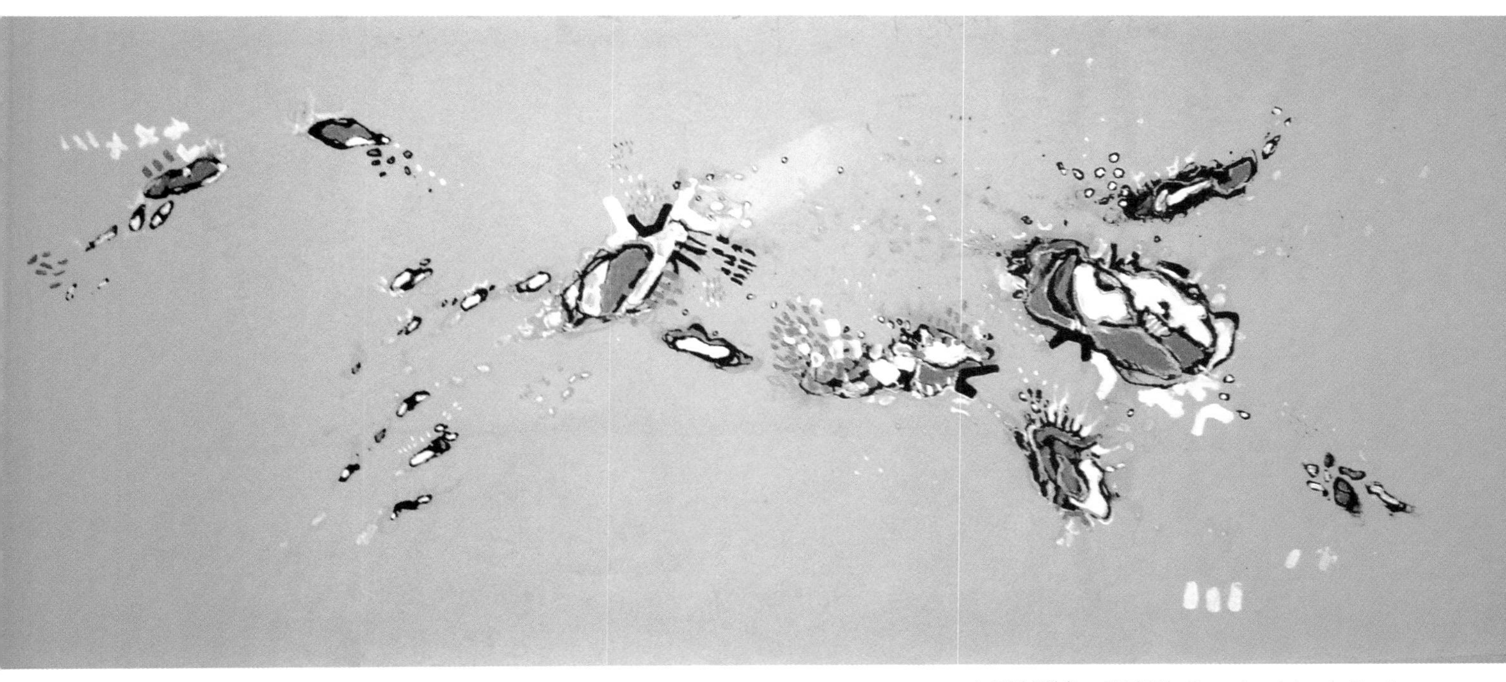

加那利群岛，西班牙 Canarian Island, Spain

*岛屿是景观的点睛之笔，不同寻常的足迹遍布独一无二的瞬间。加那利群岛的历史、故事与传说数不胜数。兰萨罗特岛坐拥完美的抽象景观，它的农业形态是当地人民生活的本质所在。*

*The islands are points of landscape, unique moments that sign unusual directions. The Canarias Islands are full of history, stories and legends. Lanzarote Island is the perfect declaration of an abstract landscape. The shape of its agriculture is an alphabet of life.*

加那利群岛，西班牙 Canarian Island, Spain

墨西拿海峡，意大利

The Strait of Messina, Italy

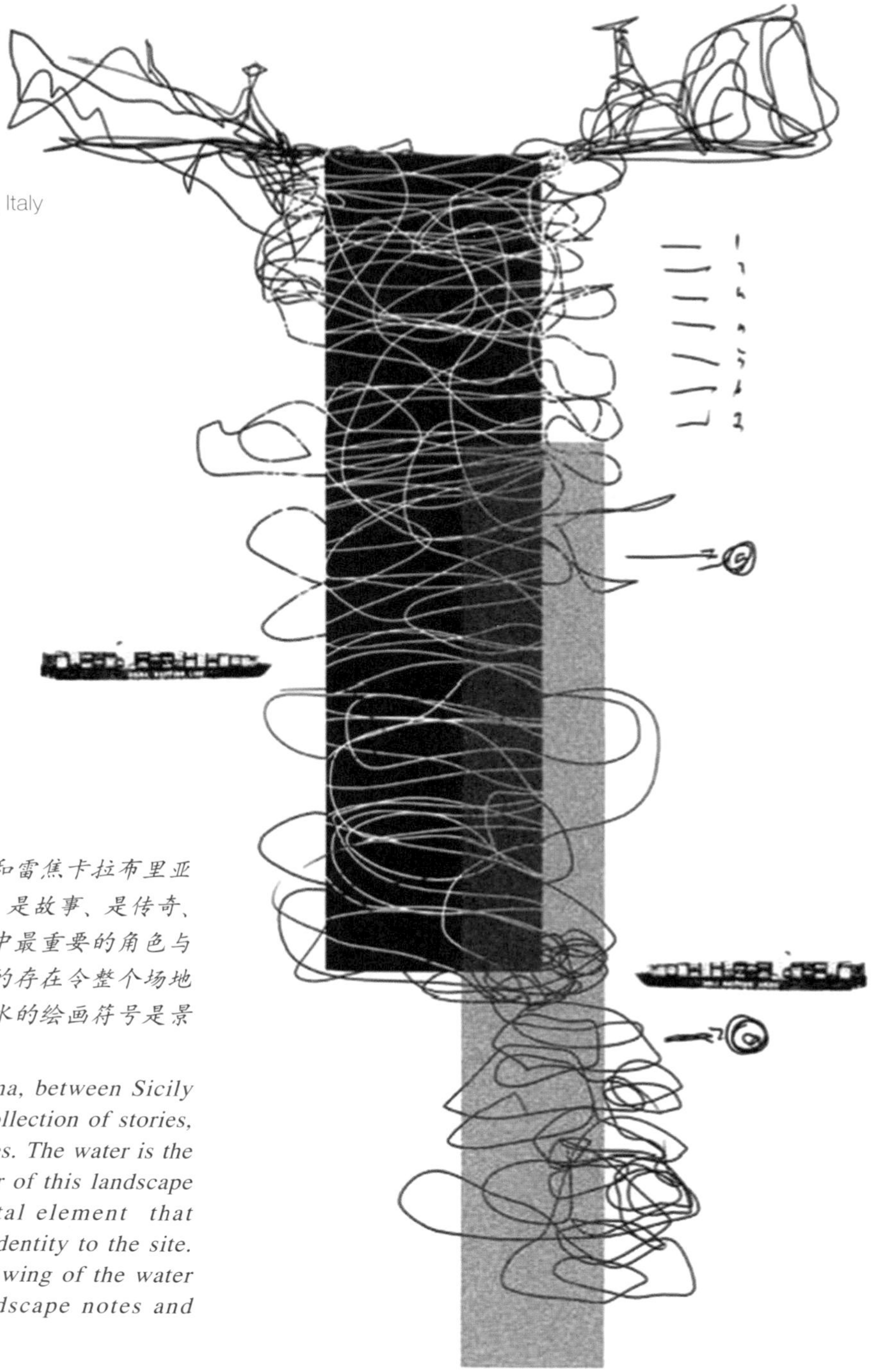

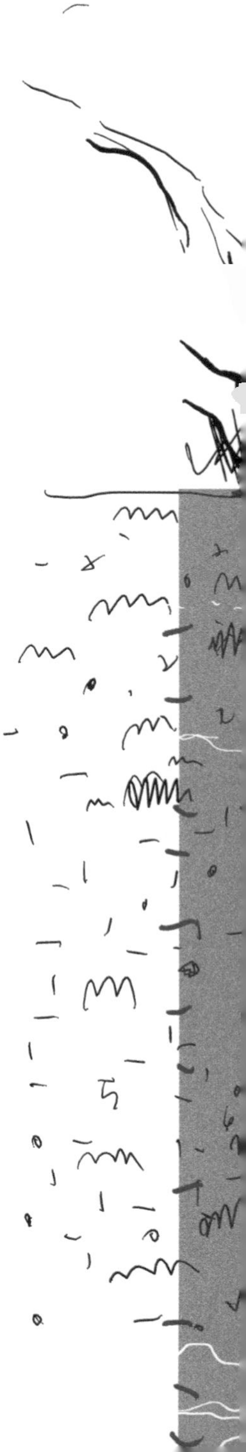

位于西西里岛和雷焦卡拉布里亚之间的墨西拿海峡，是故事、是传奇、是记忆。水是景观中最重要的角色与最基本的元素，它的存在令整个场地的特性更加鲜明。水的绘画符号是景观与记忆的标志。

*The Strait of Messina, between Sicily and Calabria, is a collection of stories, legends and memories. The water is the most important actor of this landscape and the fundamental element that gives strength and identity to the site. The signs of the drawing of the water are marks and landscape notes and memories.*

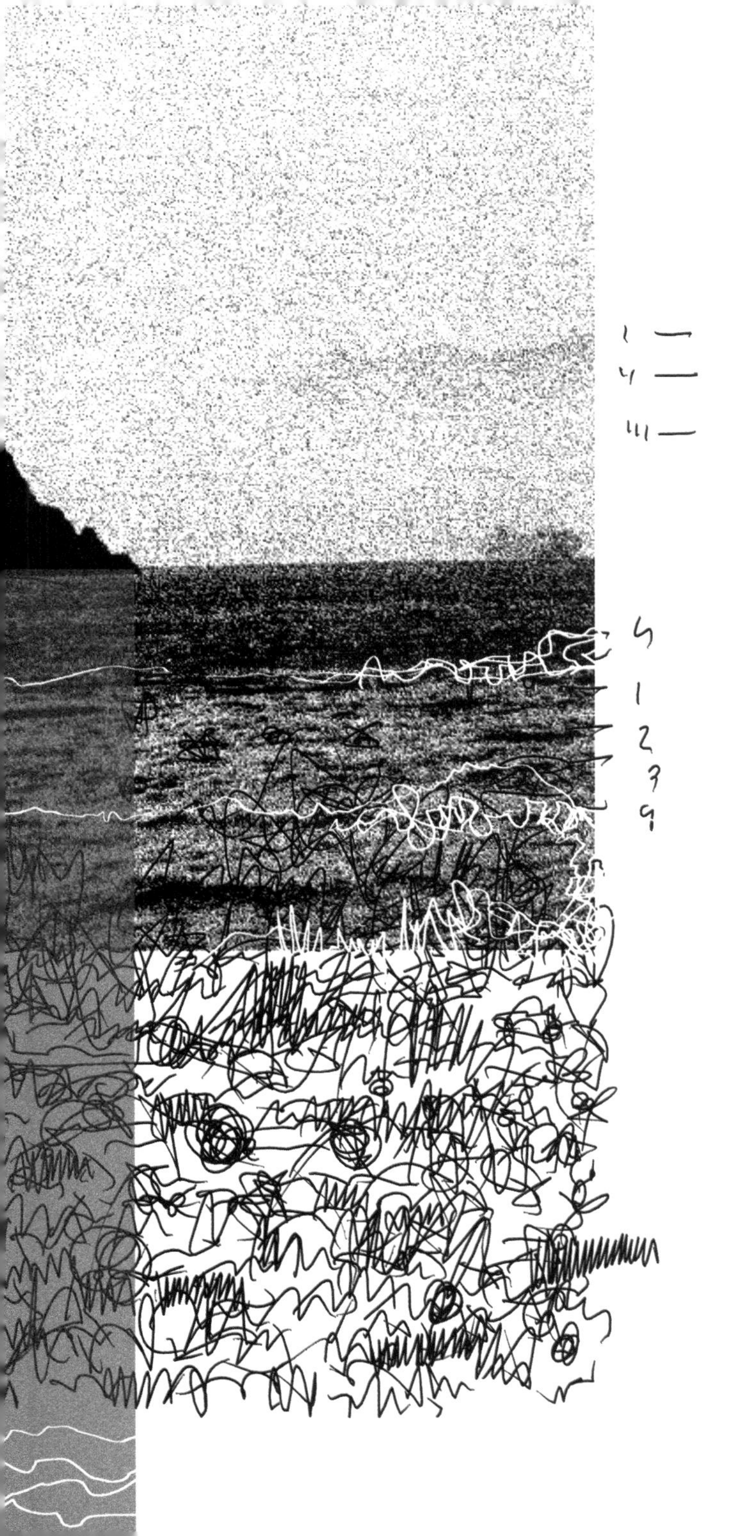

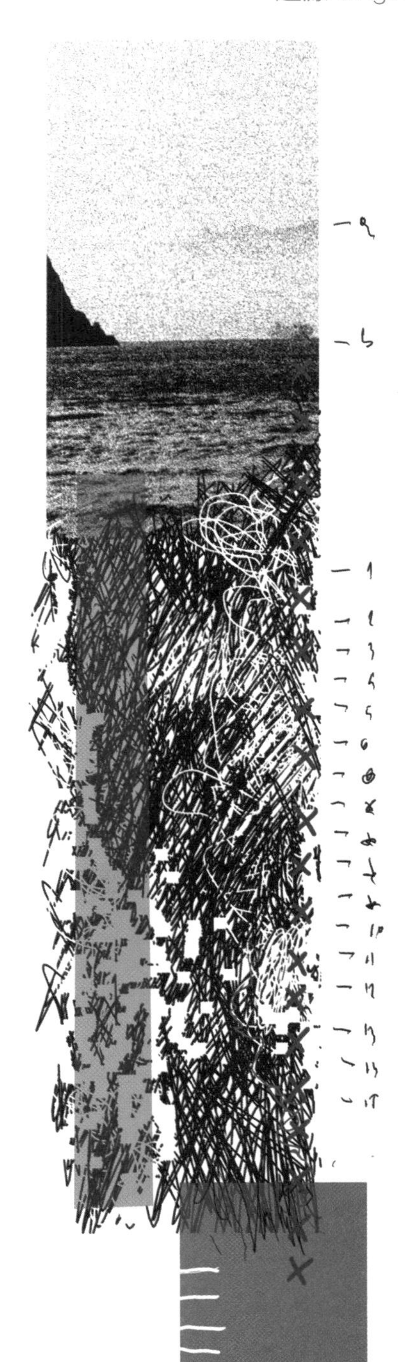

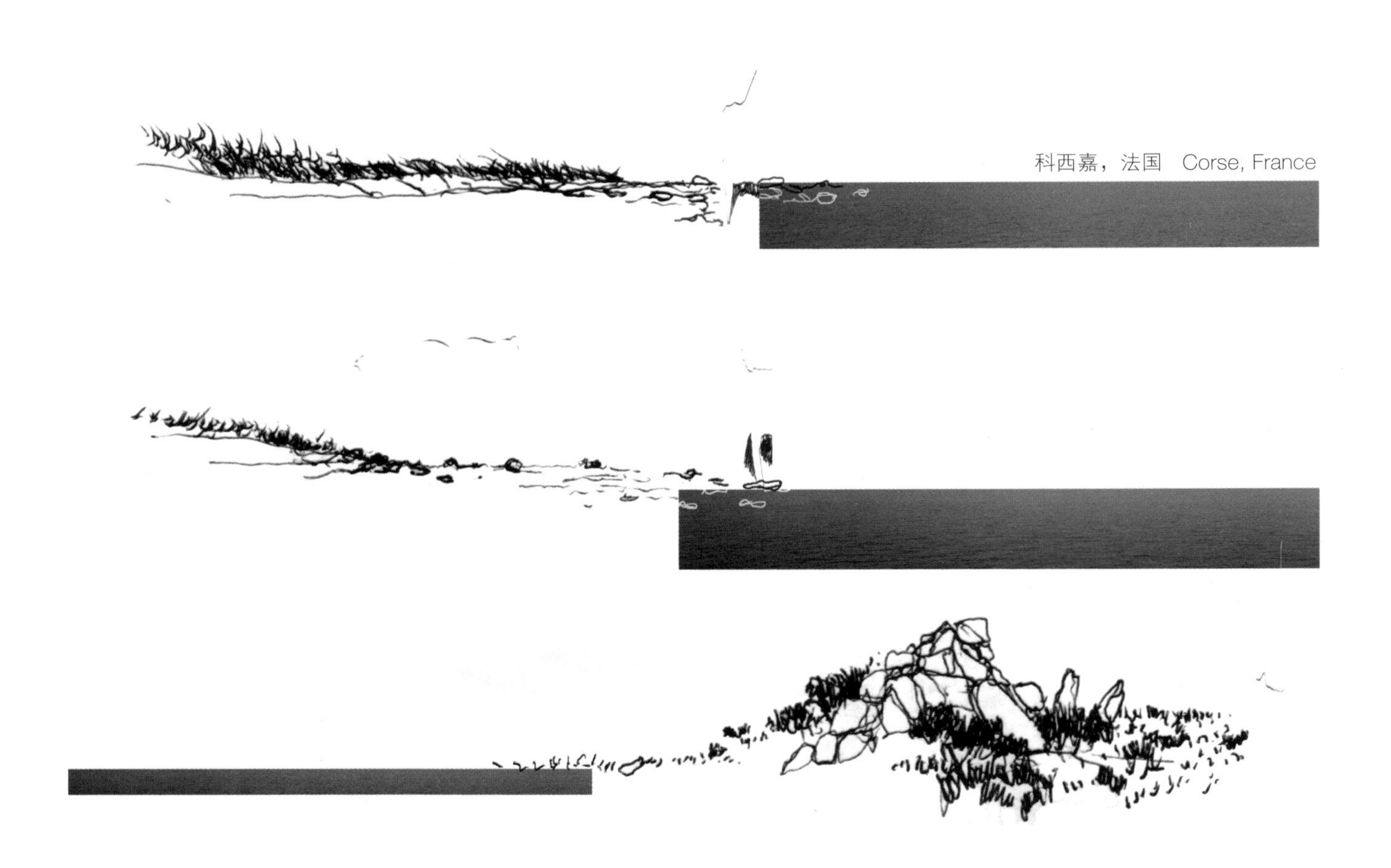


*科西嘉是景观胜地。凭借大海，山石与植被争相成为景观的主角，在无尽的天色中，相映成趣。*

*Corse is a site of landscape. The stones and the vegetation fight for the supremacy of the landscape. The sea is what they have in common. It makes them compromise through the rule of its infinitive scale.*

*夏日，降落在卡萨布兰卡的机场，摩洛哥的景观是平的；富有农耕气息的棕色渐渐弥漫开来，潜藏的光影充满着无限的不羁。颜色与线条之间的关系分明呈现了出来，它们标记且划分了阴影，农业生产规则背后不为人知的秘密亦渐渐呈现。*

*Landing in Casablanca's airport in the summer, the Moroccan landscape is flat; infinite and undefined made of shades of brown, the agricultural brown. I represented the relationship between the colors and the lines that mark and divide them. I would like to represent the secret behind the rules of the agricultural making.*

摩洛哥　Morocco

突尼斯 Tunisia

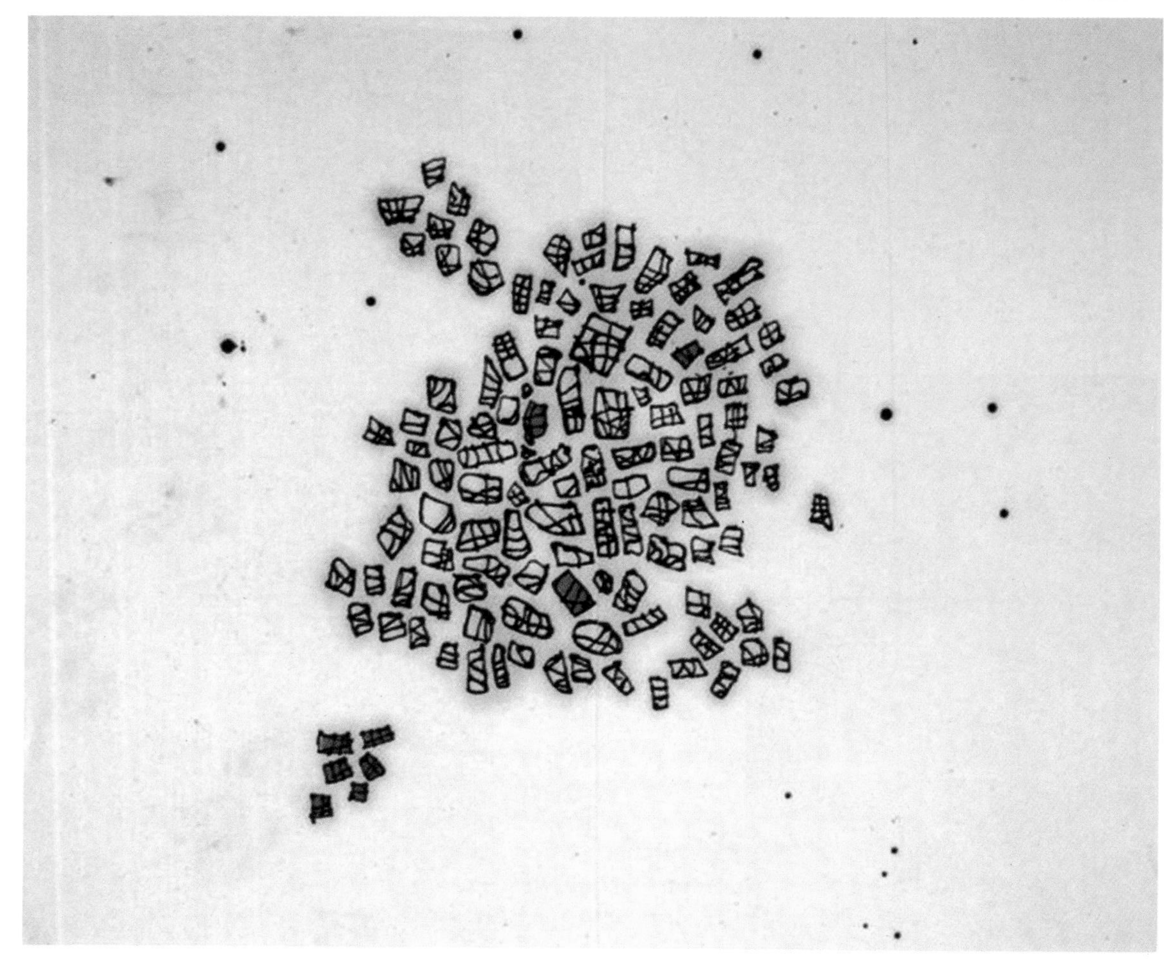

突尼斯 Tunisia

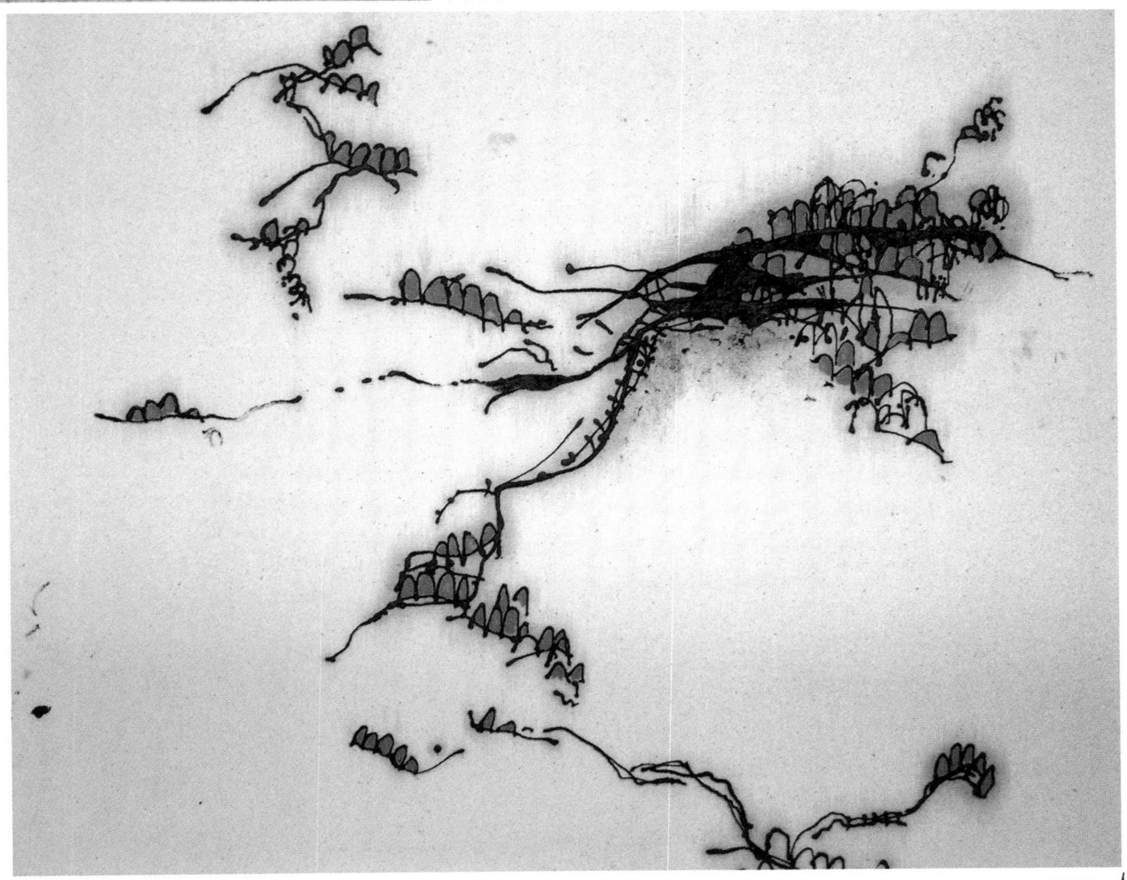

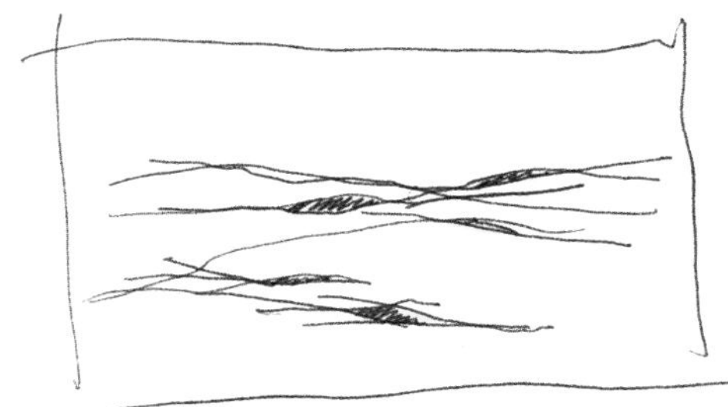

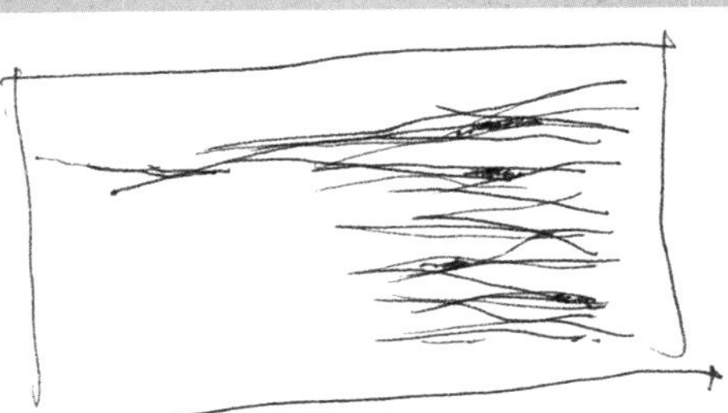

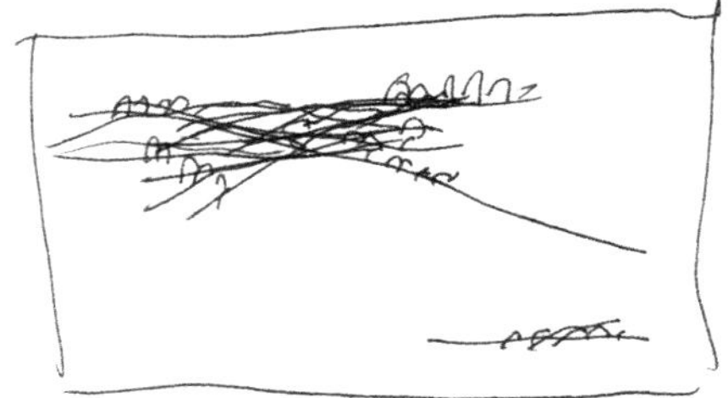

突尼斯 Tunisia

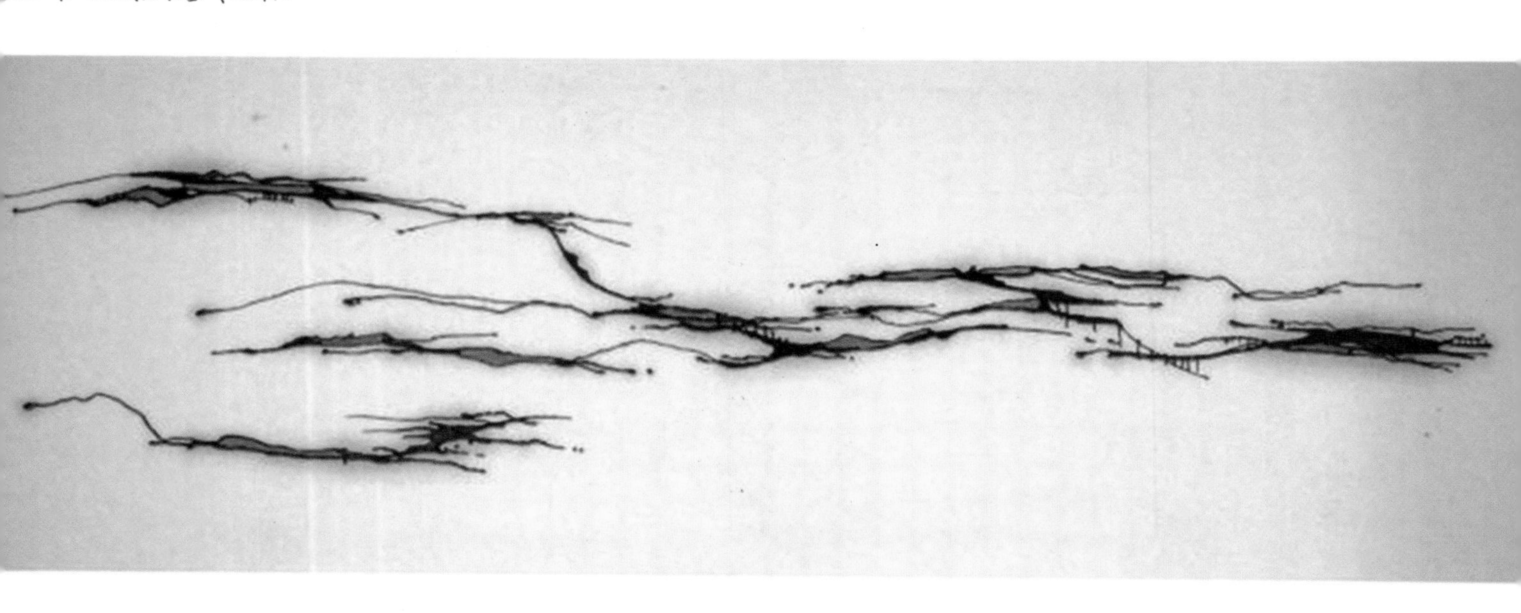

突尼斯的南部具有一种心领神会的美，每一个元素都诉说着神秘而古老的语言。沙漠形成抽象的景观，没有生命，没有色彩，似乎已被视觉的识别所忽略。然而，悄无声息中，沙漠的景观已延绵视野之外，它将色彩与形态娓娓道来，生命的意义已超越了视觉的界限。生命存在于层层叠叠的地形水平线之间，城市亦生长于此。景观延展于不同的空间，色彩也渐渐丰富起来。

*Southern Tunisia is the place of empathy where each element speaks a secret ancient language. The desert draws the landscape in an abstract way. It forces it to be invisible to itself: without life and color. However the landscape of the desert, silently, goes beyond the visible, and tells us about colors and shapes, it lets us know that life is beyond the visible. The horizontal lines of the topography creates spaces between them where it is possible to have life. Inside of these spaces the cities developed, and so the drawing of the landscape divides the space to allow the color to come in.*

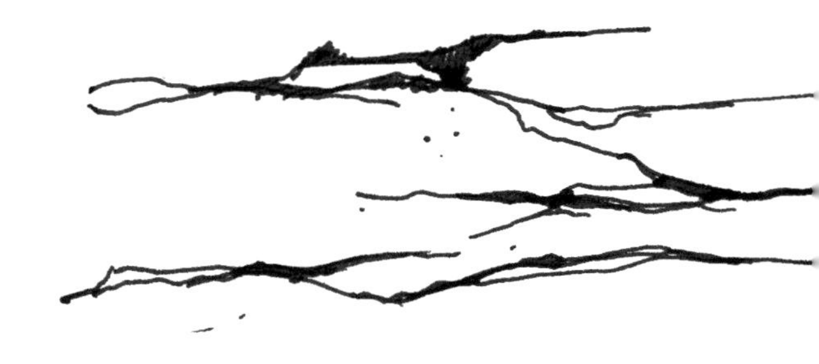

123456

# 城市中的记忆

# Memory in the City

与自然景观相比，城市这一人造空间环境，以其复杂的要素和系统关系与之大为不同。 用延伸的理念来表现城市景观，会有怎样的不同？瓦莱里奥通常在不同的尺度下进行绘画：从大尺度上，便于观察到城市全景及其与周边的关系，从细节上可能解读构成城市自身的要素关系。当你看完他关于威尼斯、拉古萨和纽约等城市的绘画时，就会明白。即使像在城市这样每件事物都被相对完美地组织和确定的情况下，瓦莱里奥仍然强调使用记忆中的信息和利用延伸观念来放大特征的理念，以便创造新的城市愿景，寻找将来改变的可能性。建筑实体照片、手绘草图、尺寸标记、剪辑、拼贴、留白等材料和方法的综合运用，构成特殊的城市景观意味和序列。

Compared to the natural landscape, cities, as the man-made environment, are different from the natural landscape in terms of the complicated relationships between diverse elements and systems. If the concept of extension is used to express city landscape, what will the difference be? Valerio often draws in different scales: from the big scale where it is possible to perceive the urban overview and its relationship with urban surroundings; to the detailed perspective where it is possible to read the relationships between the elements that compose the city itself. When you see his drawings about Venice, Ragusa Ibla and New York city, you will then understand. Even though everything is well organized and fixed, the idea of using the information from the memory and employing views of extension in order to amplify the features, is also emphasized. The ultimate aim is to create visions of new cities and search for the possibility of a future modification. The integrated use of materials and method such as photos of buildings, freehand sketches, dimensional signs, photo editing, collages, blank spaces, etc., constitutes a special sense of urban landscape and sequence.

*意大利的费拉拉被尚存的古城墙包围着。到达那里，我仿佛被敌军围攻着。片刻后，我开始表现这支军队的精确组织，它的规则和关系，由需求和战斗策略组成。*

*The city of Ferrara, in Italy, is surrounded by ancient walls that still have not been touched. Getting there, it is as if I imagined a siege of an enemy army. After a while I represented a precise organization of this army with its rules and relationships consisting in demands and battle strategies.*

费拉拉，意大利　Ferrara, Italy

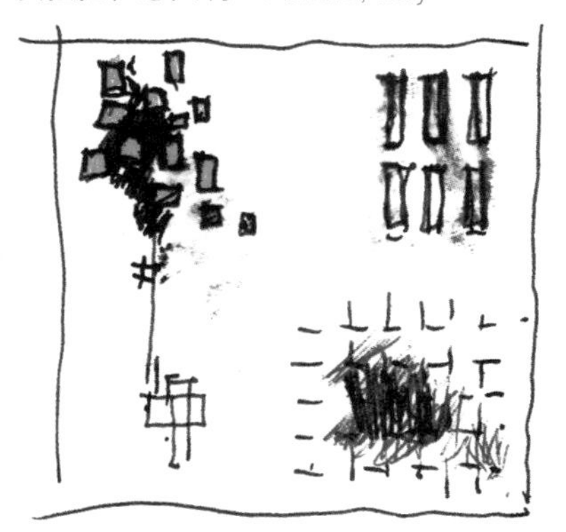

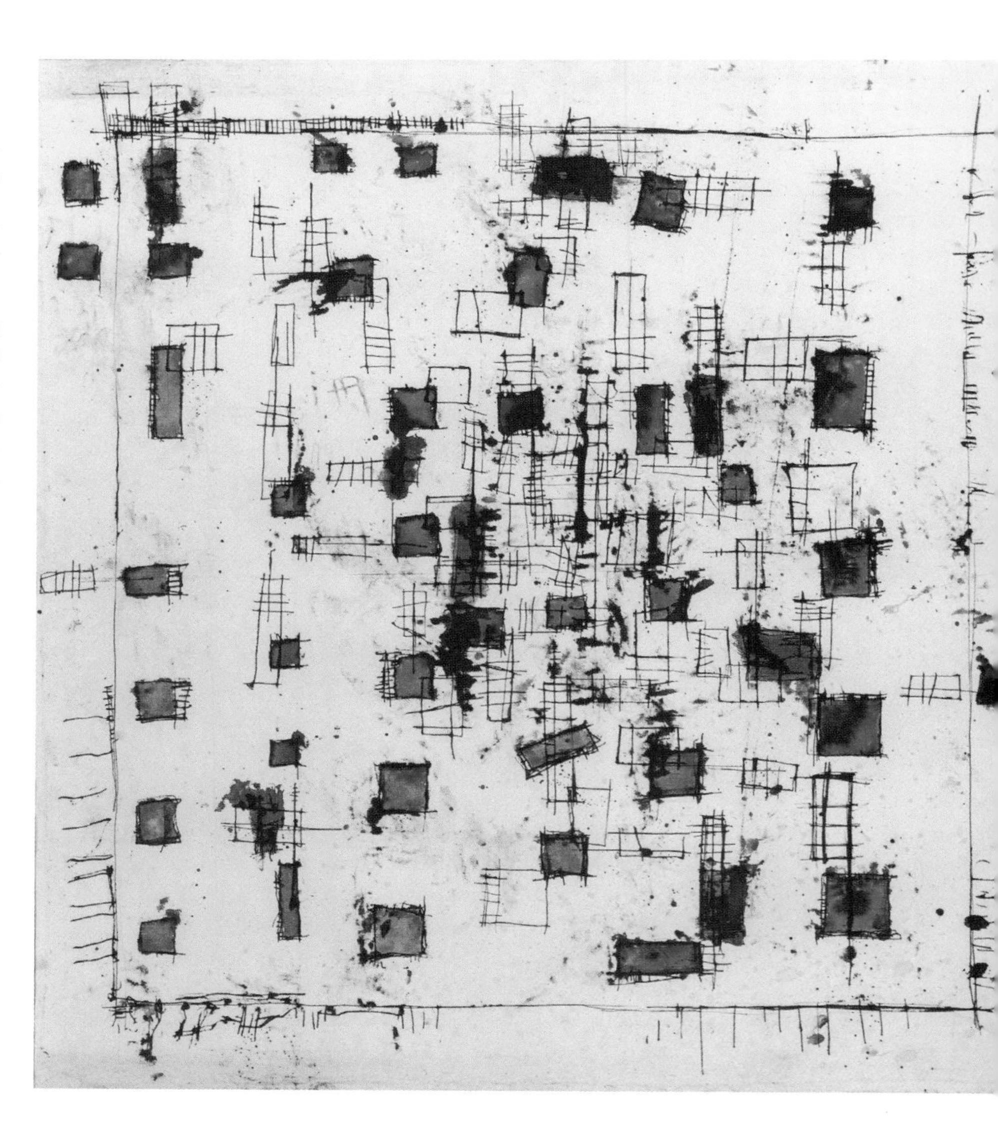

威尼斯，意大利　Venezia, Italy

威尼斯是一座水城，水与城的关系，被转译了，通过向水面渐变的红色街区的精确组织，中间的空白则为未来的想法及梦想预留着。

*The city of Venice is the city of water. I translated this relationship through a precise organization of red blocks fading into the water, with a remaining middle space in which it is possible to add ideas and dreams.*

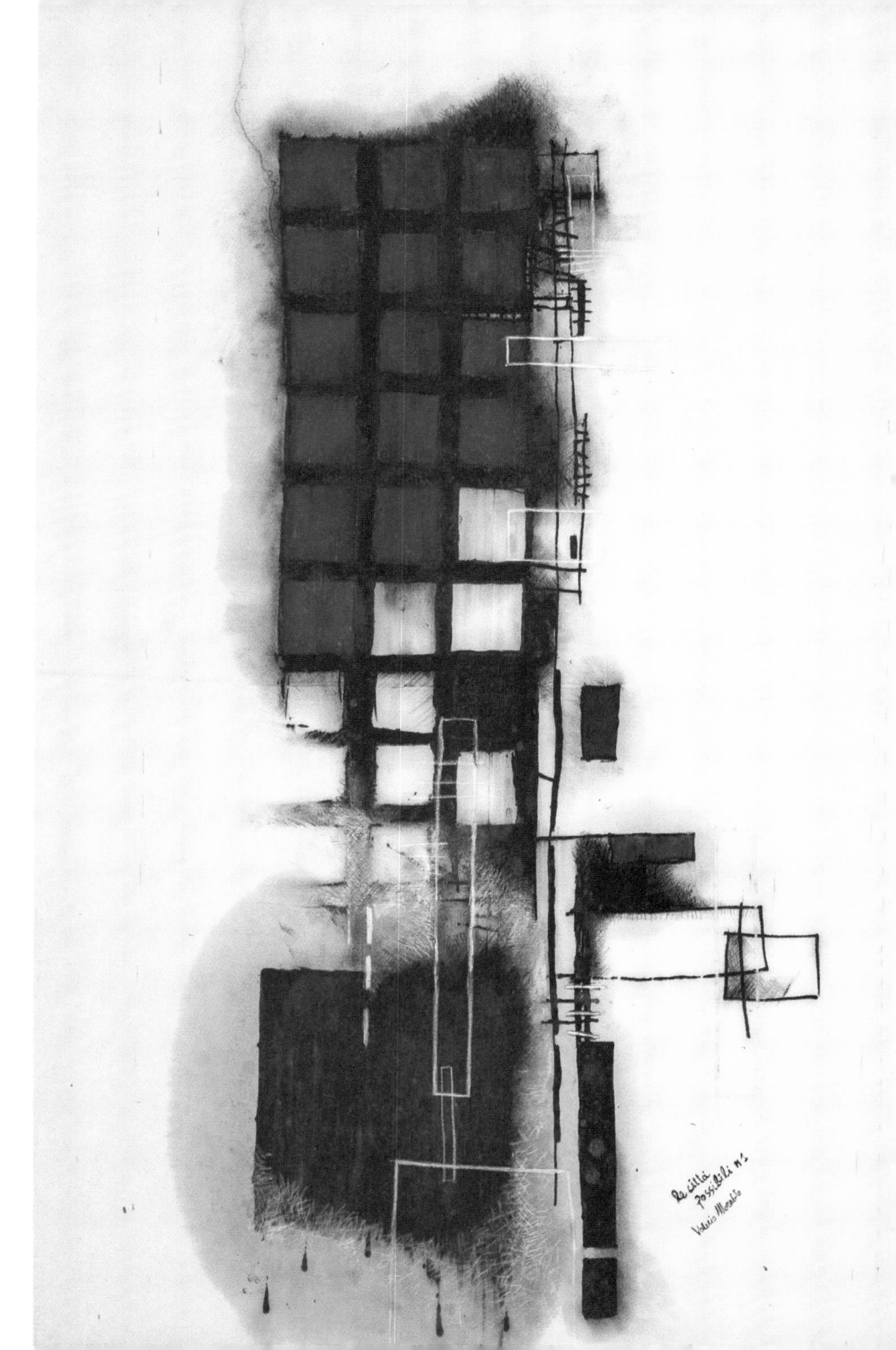

*有时没有必要为了获得特殊的感觉而在一个地方停留，快速地通过并留下某种印象即足够了。途经杜伊斯堡至奥尔登堡的高速路时，我看到了由烟囱组成的大工业区，生锈的体块和楼梯。当我觉得要画它时，我想创造一种可以视为宝藏的自然。*

*Sometimes it is not necessary to stay in a place in order to receive a particular feeling, but it is sufficient to pass quickly through it to keep in mind certain visions. Going along a highway, passing through the Duisburg forest up to Oldenburg, I believe I saw a large industrial zone made by chimneys, rusted volumes and staircases. When I decided to draw about it, I wanted to create a kind of nature that could behold a treasure.*

杜伊斯堡，德国
Duisburg, Germany

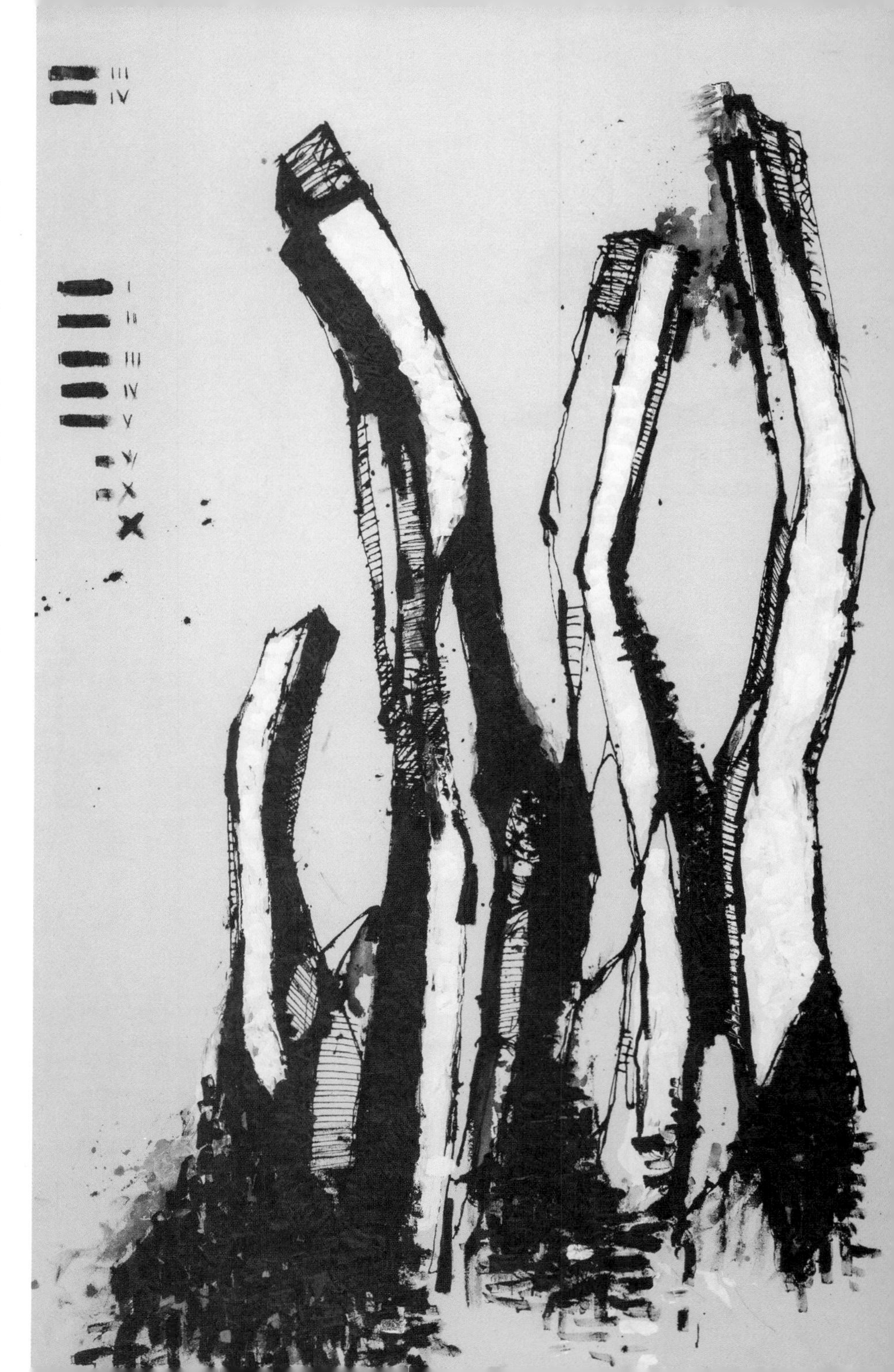

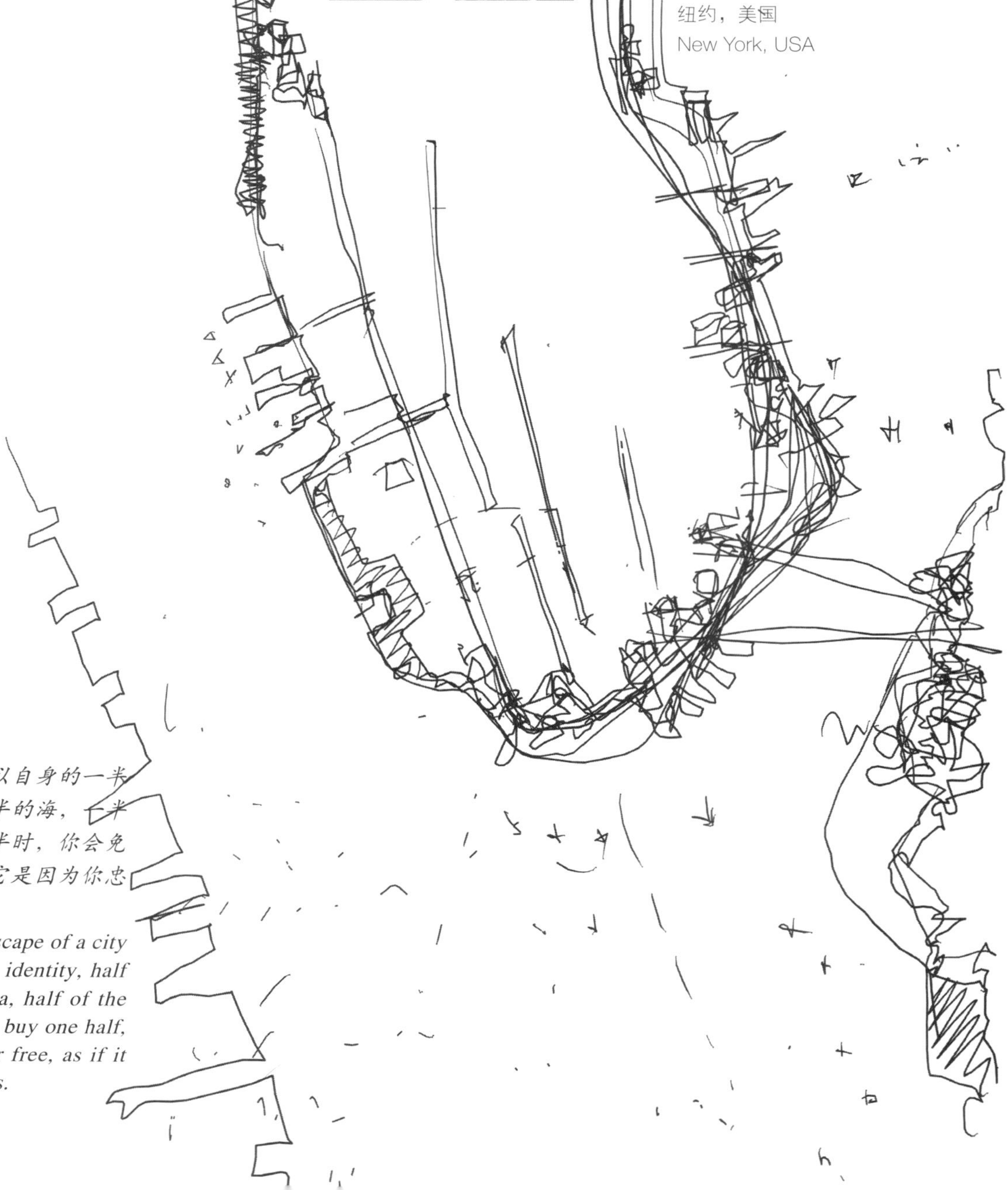

纽约，美国
New York, USA

*你会发觉城市景观以自身的一半为特性，一半的天，一半的海，一半的绿；每次你“买”一半时，你会免费获得另一半，就好像它是因为你忠诚而给予你的奖励。*

*You would percept a landscape of a city which has its own half as identity, half of the sky, half of the sea, half of the green; and every time you buy one half, the other half is given for free, as if it were a prize to faithfulness.*

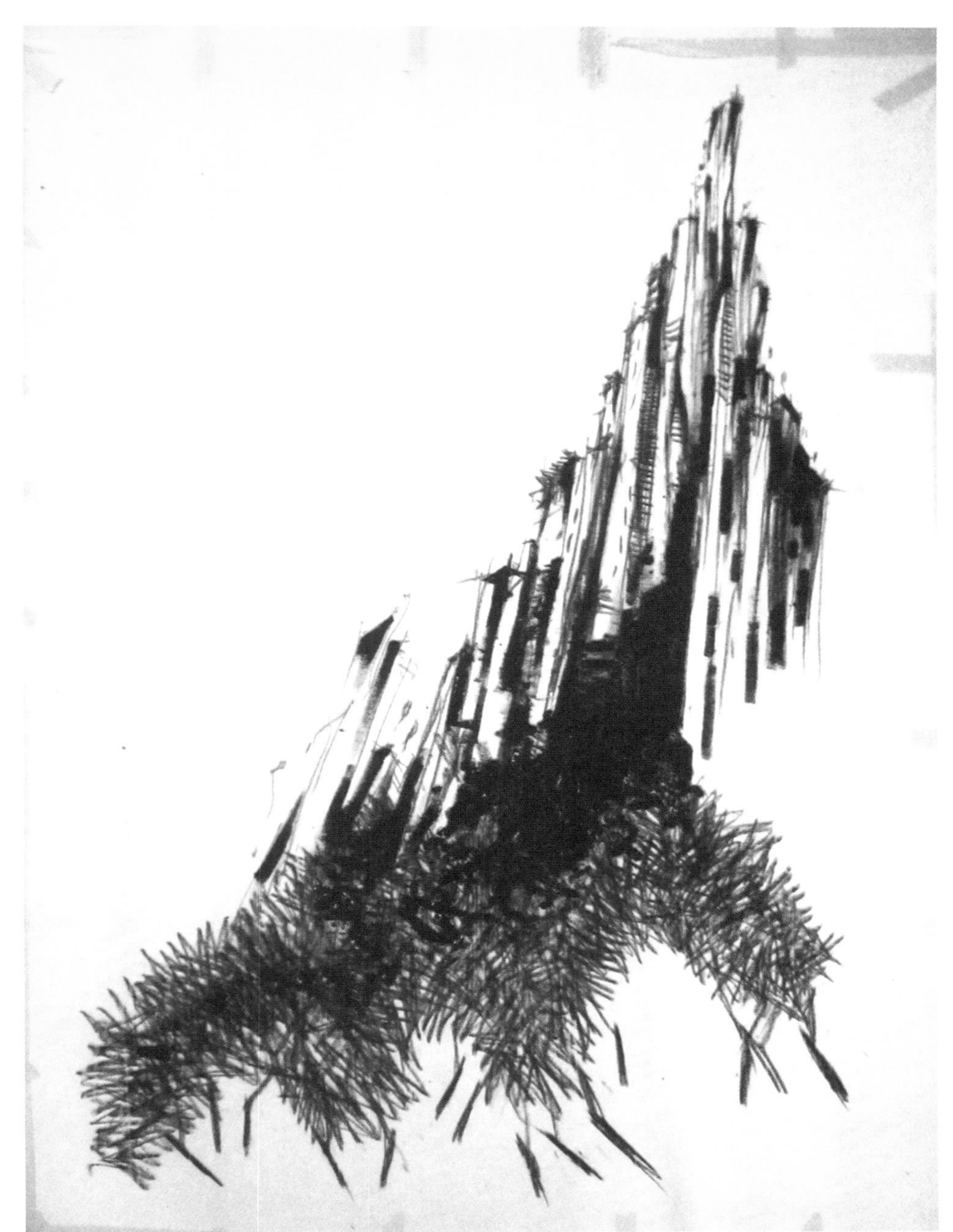

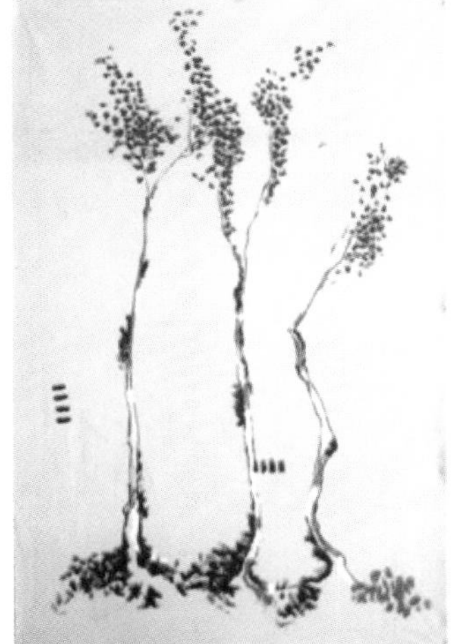

纽约，美国
New York, USA

*每一条街道似乎都透露着事物的秘密，每一条街道似乎都凝聚着事物的魔力，而魔力也维系着街道。如果街道有什么东西是与空间相对立的，那就是空隙，它们能够揭示取消的神秘欲望。*

*Every street seems to reveal the secret of things, every street seems to retain the magic of things, and magic retains the streets. And if the streets had what is contrary to space, that is the gap, they could be able to reveal the arcane desire of cancellation.*

纽约，美国
New York, USA

纽约，美国
New York, USA

西西里巴洛克风盛行于拉古萨，它是拉古萨的标志所在。描绘拉古萨是一段对秘密程式想象和创造的练习，它能够打开神秘的世界，它是一个我试图偷走且重述的巴洛克秘则。所以，在拉古萨，写满巴洛克规则的墙被重述了，它变成了童话、传说和故事，那样轻盈，以至于失去了装饰的重量。

*The city of Ragusa is a city in which the Sicilian Baroque is exuberating and iconic. Drawing Ragusa has been a stylistic exercise of imagination and invention of secret formulas, capable of opening arcane worlds. A secret code of Baroque that I tried to steal and rewrite. Therefore, the walls of Ragusa on which the Baroque codes have been written, become fairy tales, legends and stories, so light to lose the weight of the ornament.*

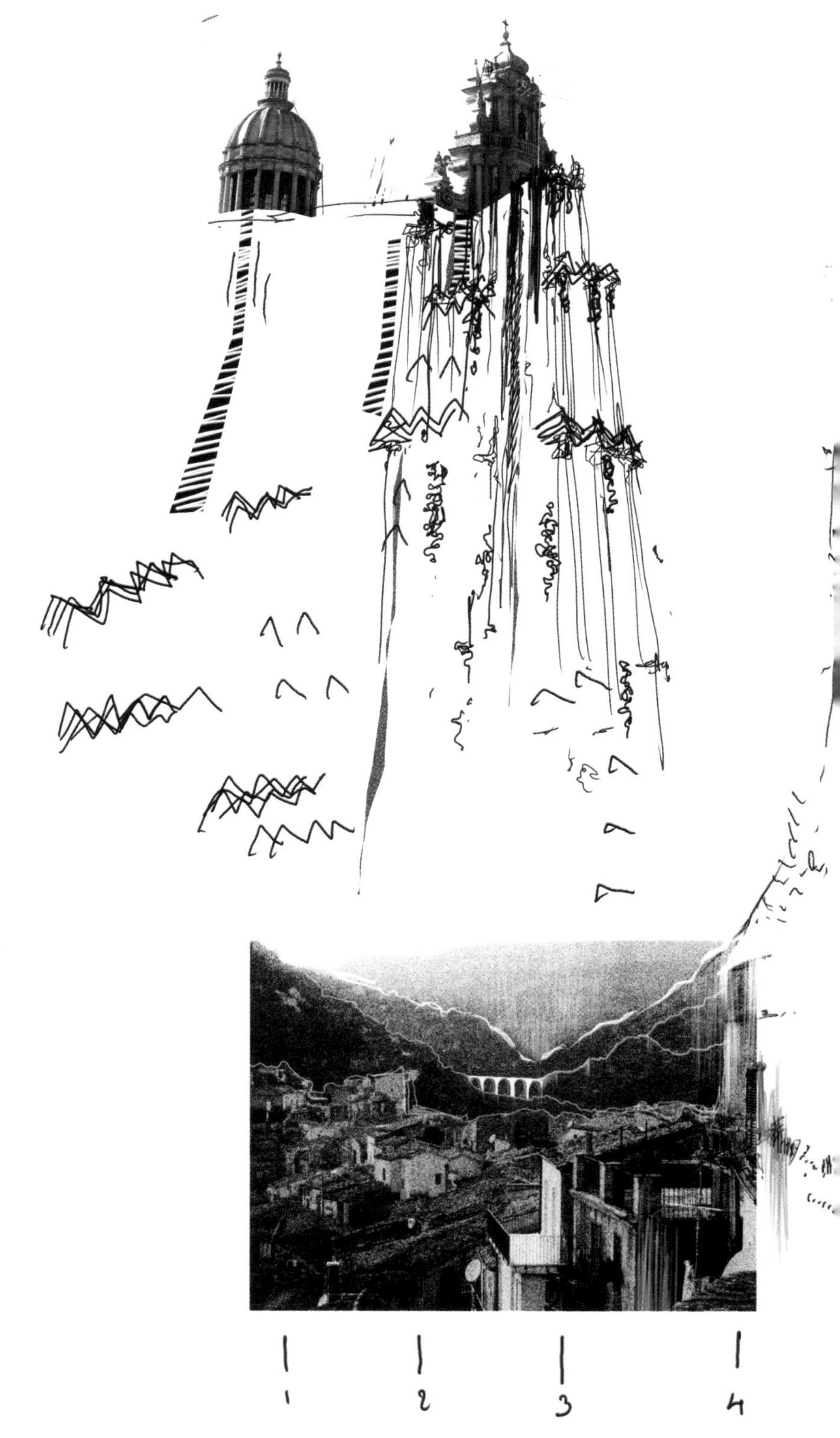

拉古萨，意大利　Ragusa Ibla, Italy

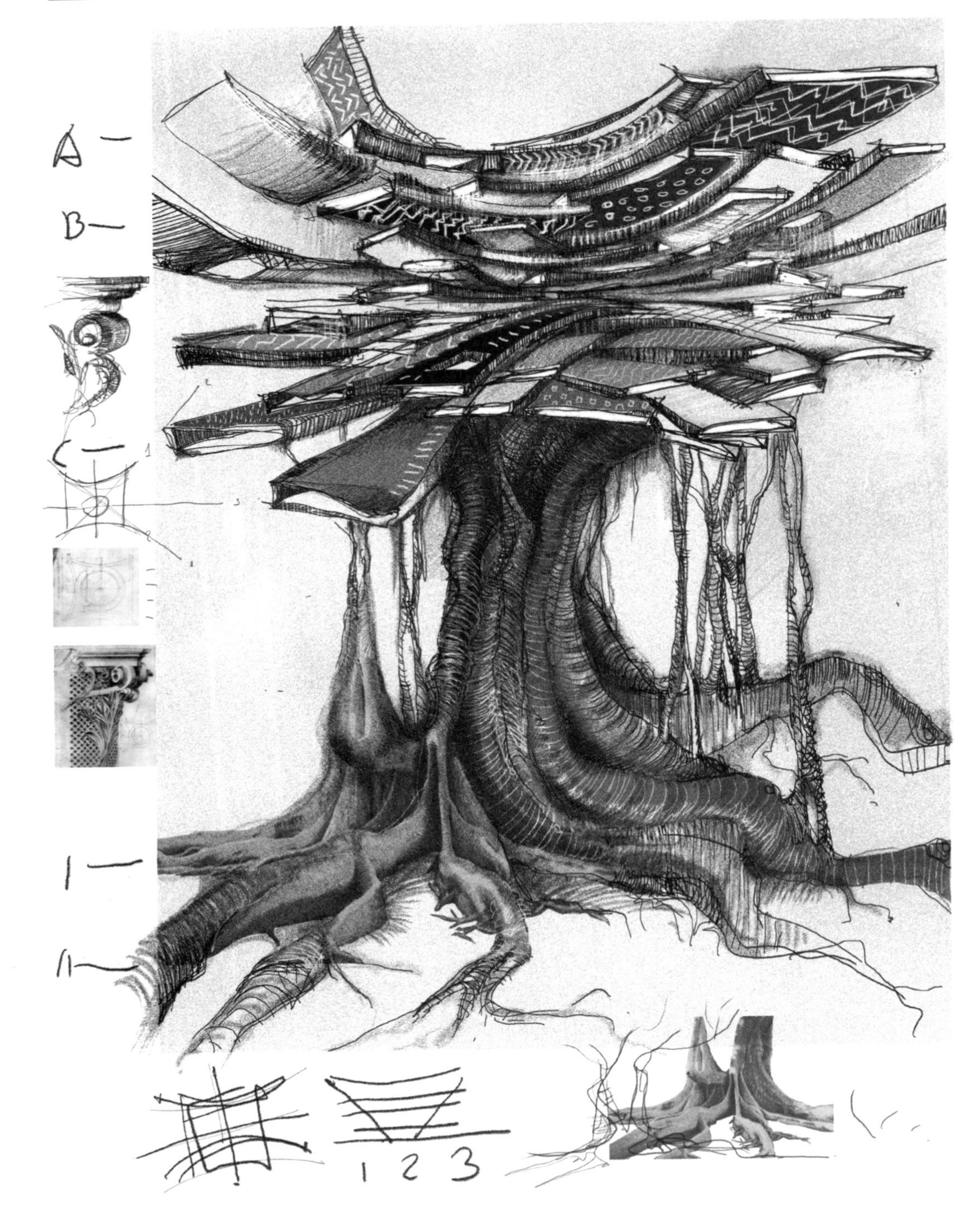
A
B
C
1 2 3

# 记忆中的记忆
# Memory in the Memory

正如我们之前提到的，记忆是延伸表现的基本要素，它存储、积累着信息，在绘画与表现时发挥作用。除了对实际存在的场地空间的表现，还有一些绘画是基于对未游览之地的印象或者对不存在之地的想象所创造的。在这一情况下，与绘画主题相关的累积的记忆信息辅助完成这种抽象的表现。瓦莱里奥在未到中国之前所创作的中国景观画就属此类典型，创作源自他对文献、文化和历史资料的阅读，对中国古地图，中国画的研习等形成的记忆。因此，可以说记忆是一种强大的资源，当你有延伸的观念时，凭记忆往往能创造一些出人意料的空间和绘画。

As mentioned earlier, memory is the fundamental element of the extended representation, in which it stores and collects the information that takes place during the process of drawing and representing. In addition to the representation towards the real place and space, there are also drawings from the memory or imagination linked with places that have never been visited or with places that do not even exist. In this case, the accumulated information in the memory, related to the topic of drawing, helps to develop the abstract representation. Typically, Valerio created Chinese landscape drawings before visiting them, and they derived from the reading of literature, culture and history, from the study of ancient maps and traditional paintings etc. Then it can be understood that the memory is a powerful resource from which innovative space and drawings may be generated if you have the idea of extended representation.

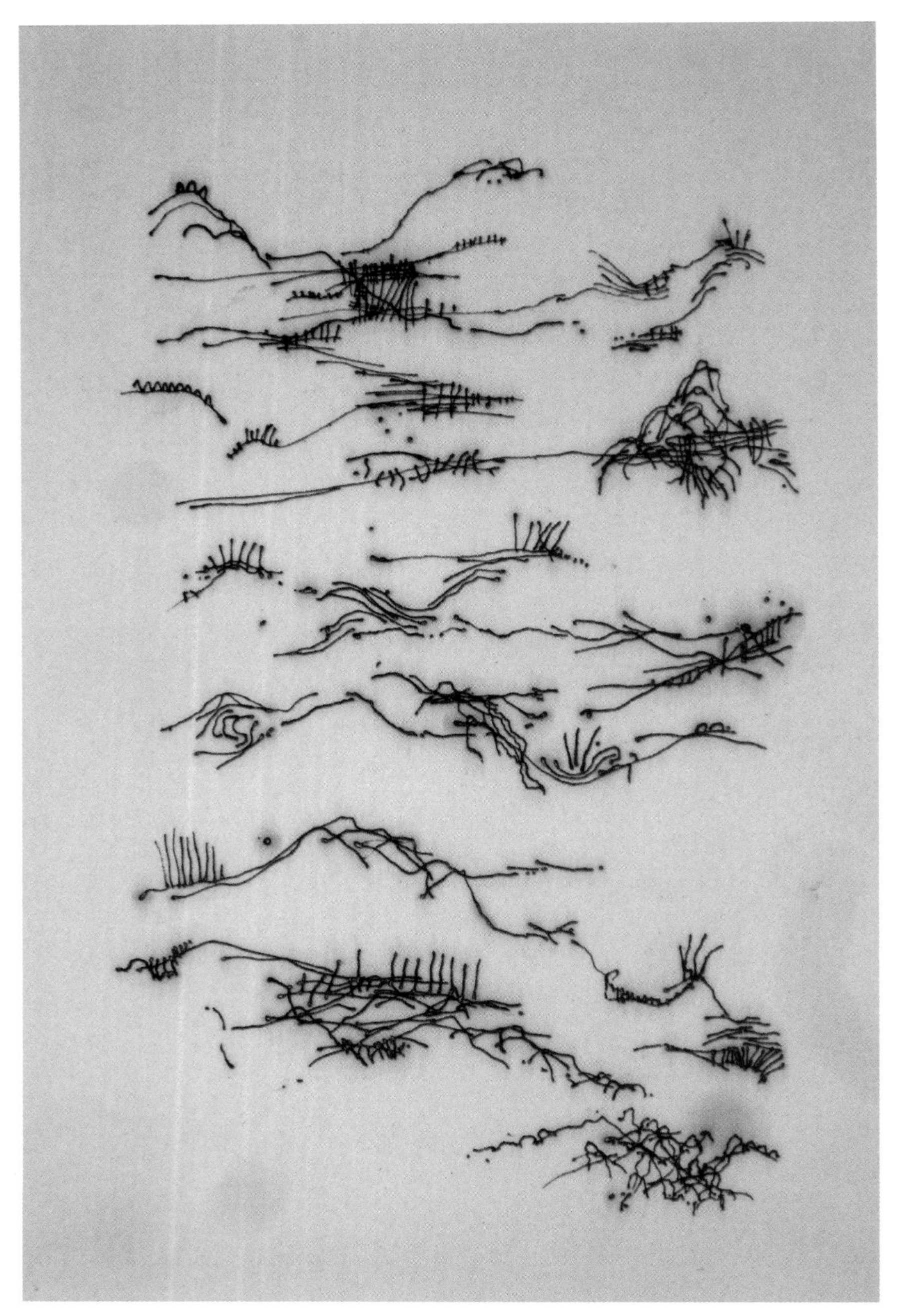

*我认为，希腊是声之土地。抽象的天际线中萦绕着蟋蟀与蝉的鸣响，而它们却对音律一无所知。在我的绘画中，试图谱写一曲乐章来描绘它们的音符。所以，我开始创造，开始寻找这种合成的微妙。*

*In my opinion, Greece is the land of sound. It is the sound of the crickets and of the cicadas that draws abstract lines in the sky: they seem as unknown musical notes. I would like to put the sound notes in an invented musical score through my drawings. So I came up with inventions and relations of synthesis.*

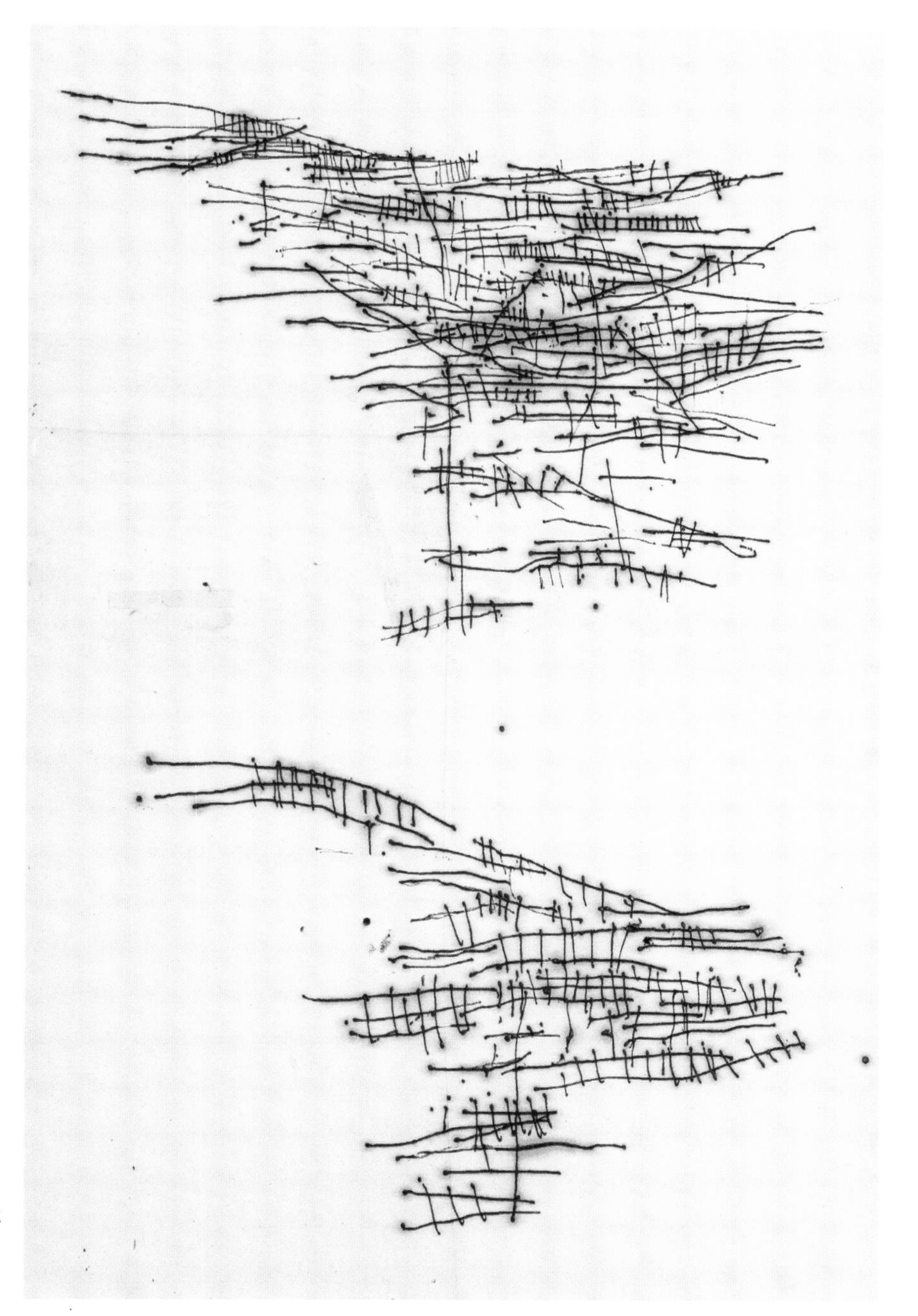

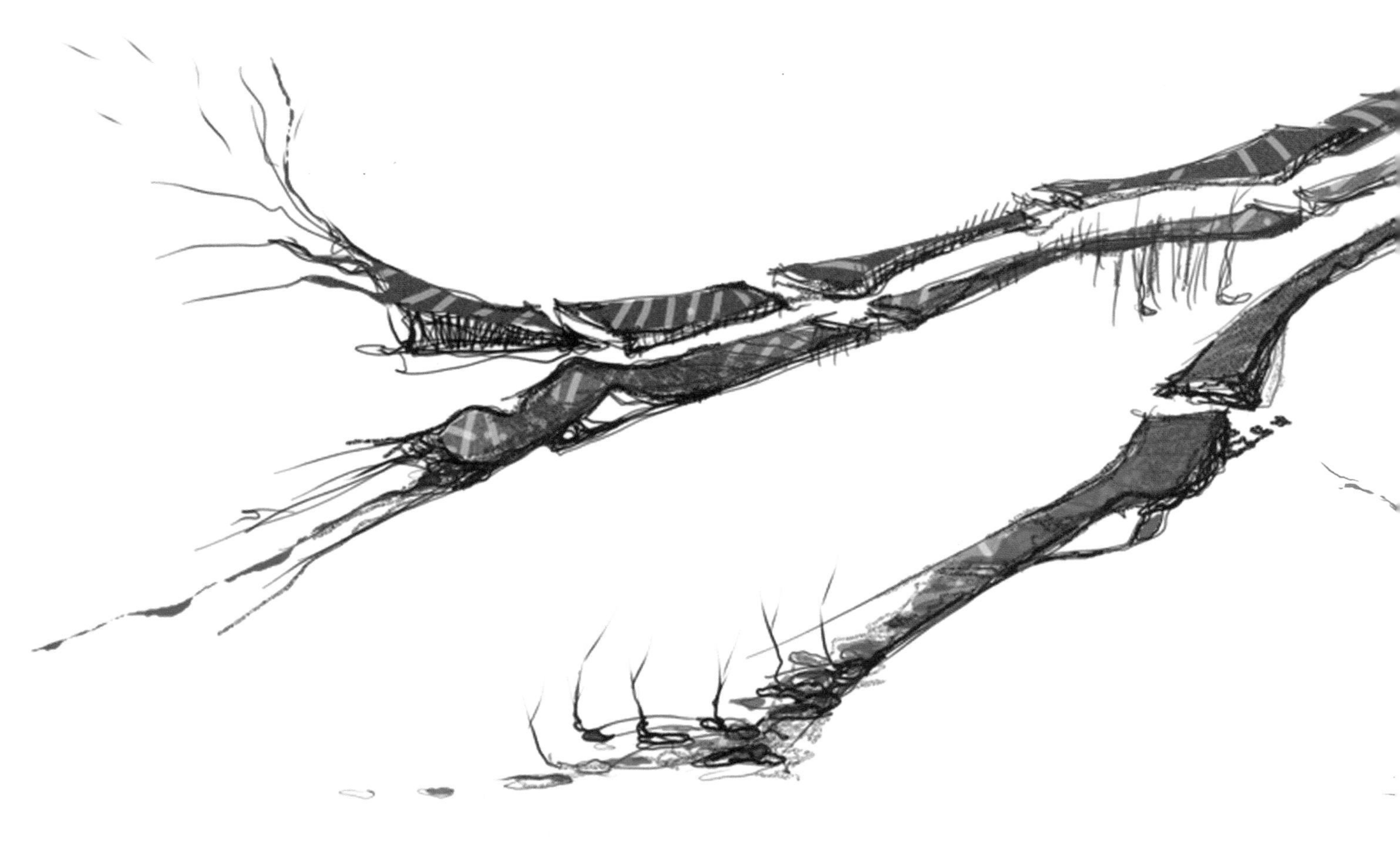

*当我画一个并不存在的地方时，它的特征源于所有我到访的地方。在那里，建筑与景观形成一种结构，遵循着相同的计划与方法，它们在绘画中密不可分。空间与材料的虚拟组合无限地伸展流动着。*

*When I draw a place that does not exist , its characteristic is made of the sum of all the places I have visited. In these places architecture and landscape get their structure following common schemes and methodologies. In these drawings the two realities are built an undividable way. The unreal combinations of the space and the materials follow endless flows.*

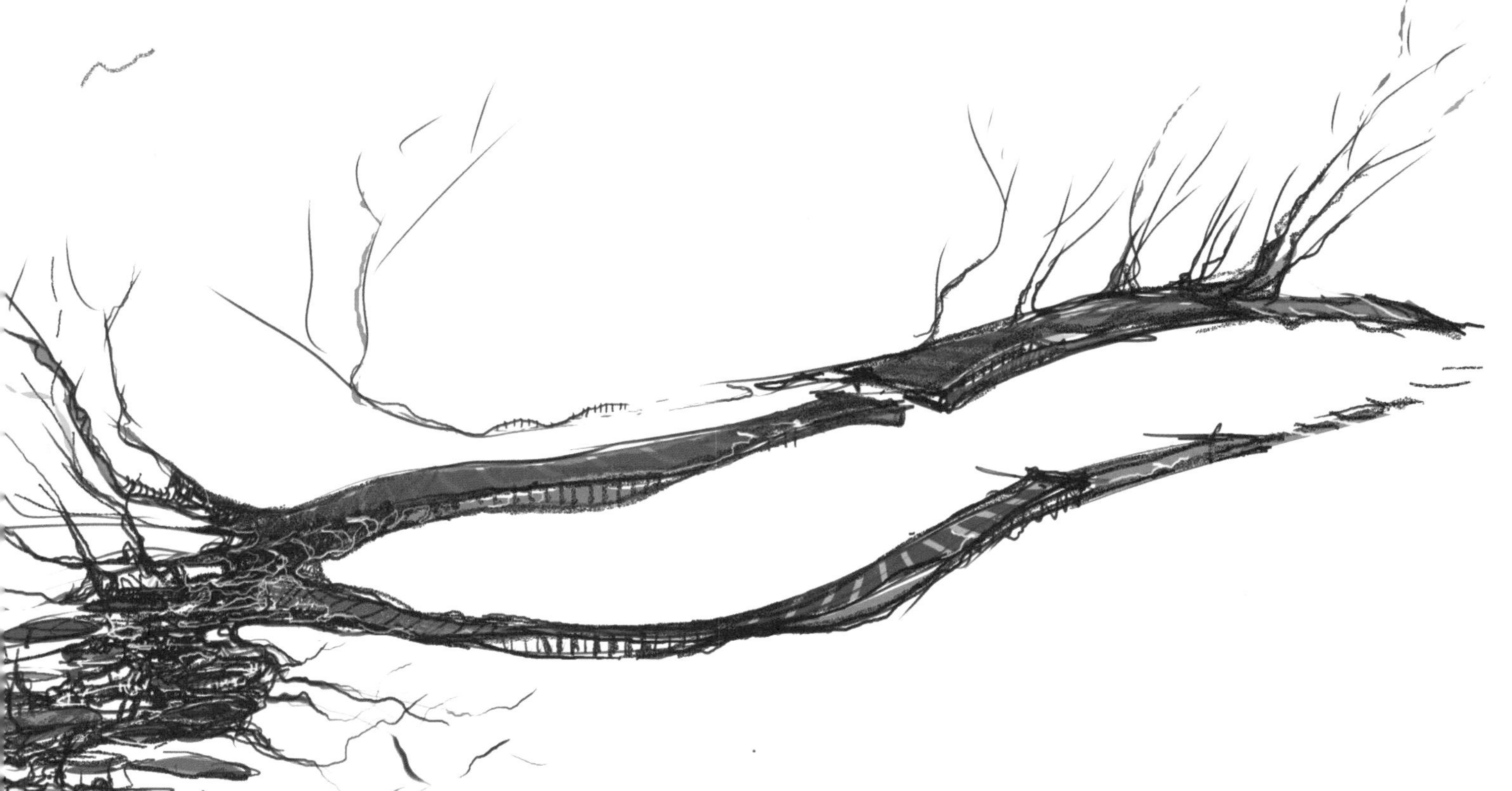

仙境景观（蒂姆·伯顿感想）
Landscape in the Wonderland(Thinking of Tim Burton)

景观与记忆
Landscape and Memory

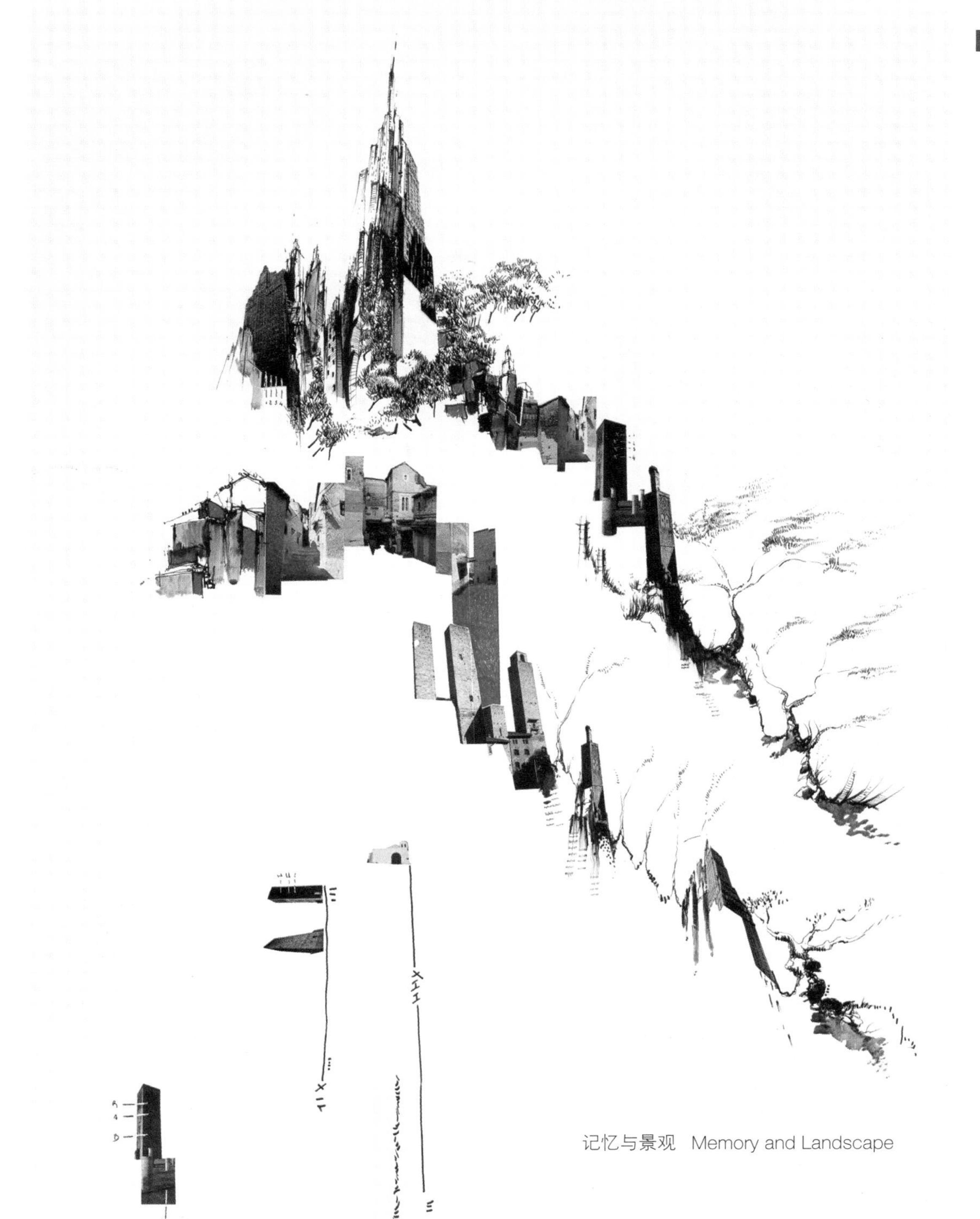

记忆与景观　Memory and Landscape

拜访雅典卫城的博物馆时，我惊讶于古代陶瓷的色彩。罐子，花瓶和其他物件都由两种颜色来表现：大地棕色和抽象黑色。这两种颜色和图纹装饰的组合效果令人难以置信。于是，我回到意大利就决定通过使用棕色，黑色和装饰的不同组合来创造新的城市意象。

*When I visited the Museum of Acropolis in Athens, I was impressed by the color of the ancient ceramics. Pots, vases, and other objects were represented by two colors: brown from the earth, and black as an abstract color. The combination between the two colors and the ornaments of the drawings let me astonished. When I went back to Italy I decided to draw new images of imaginary cities using the idea of brown, black and the combination of different ornaments.*

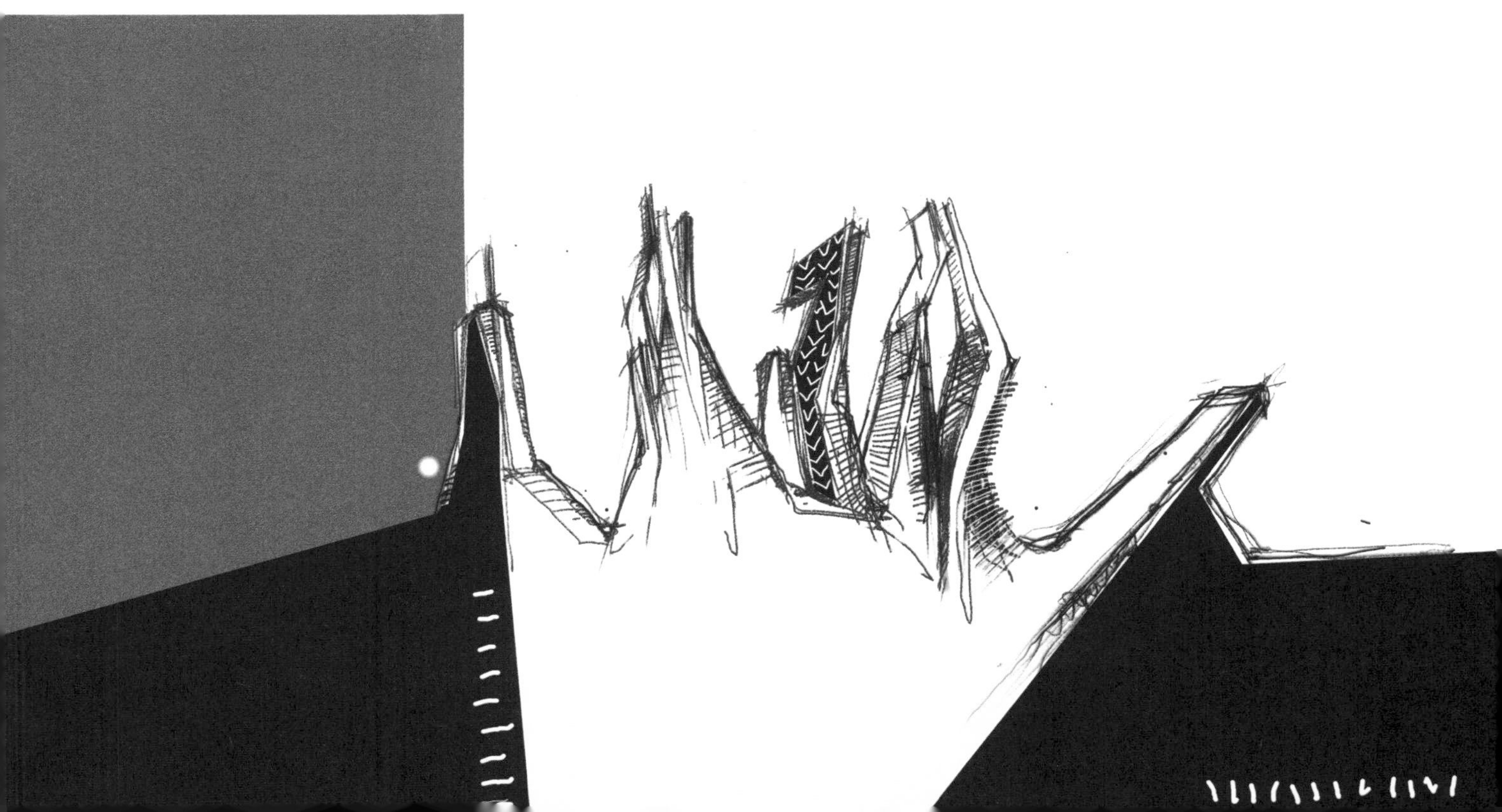

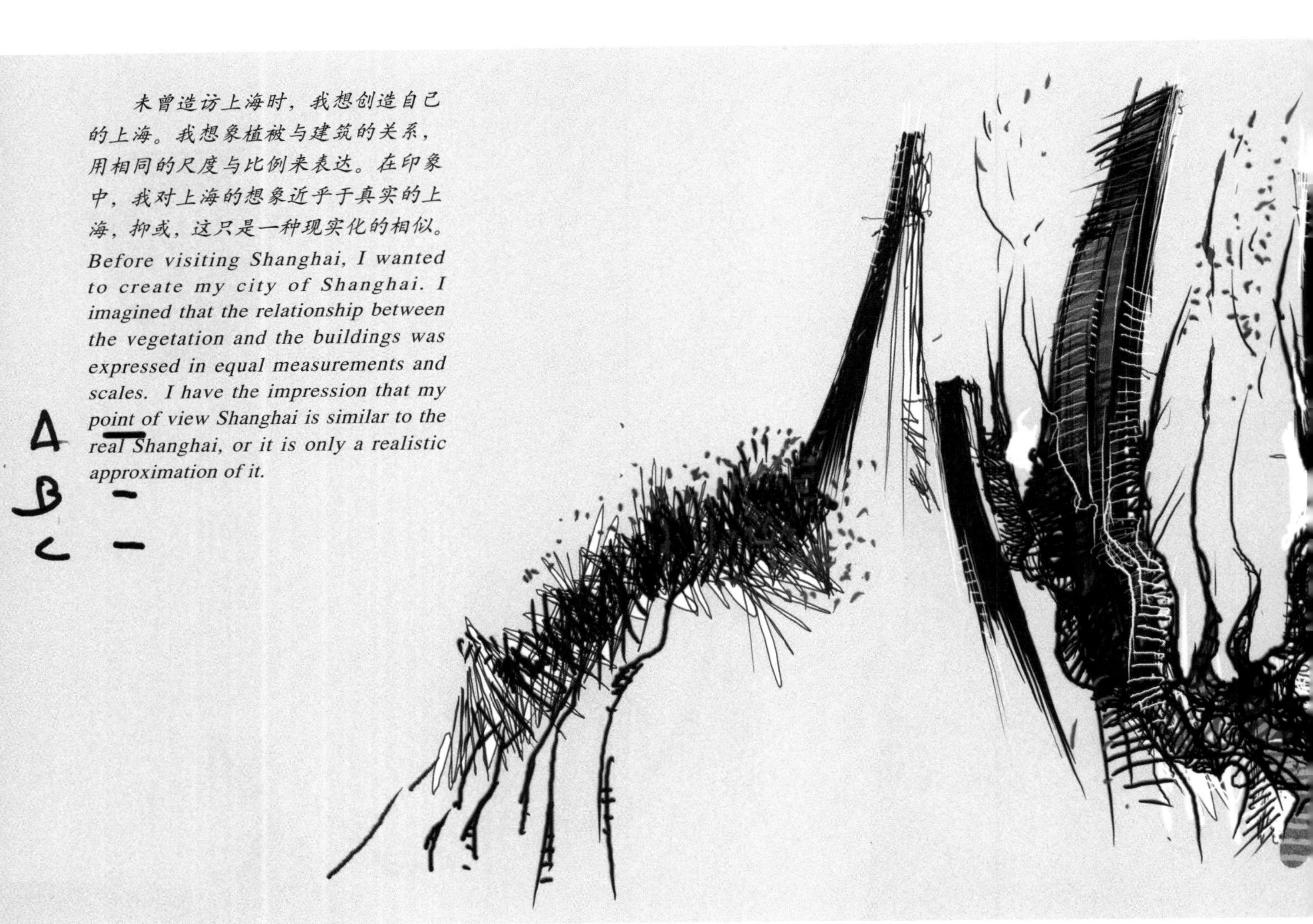

未曾造访上海时，我想创造自己的上海。我想象植被与建筑的关系，用相同的尺度与比例来表达。在印象中，我对上海的想象近乎于真实的上海，抑或，这只是一种现实化的相似。

*Before visiting Shanghai, I wanted to create my city of Shanghai. I imagined that the relationship between the vegetation and the buildings was expressed in equal measurements and scales. I have the impression that my point of view Shanghai is similar to the real Shanghai, or it is only a realistic approximation of it.*

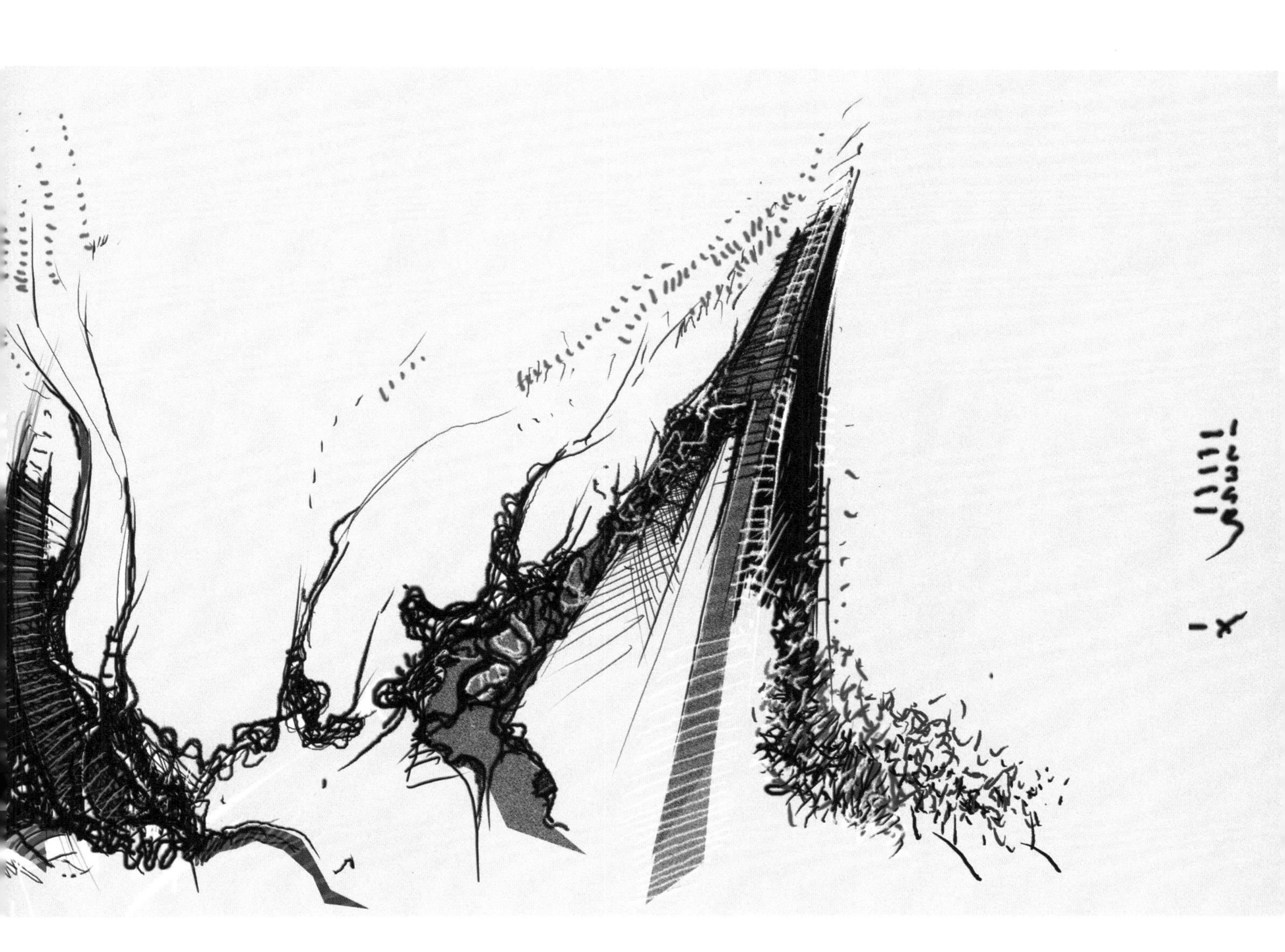

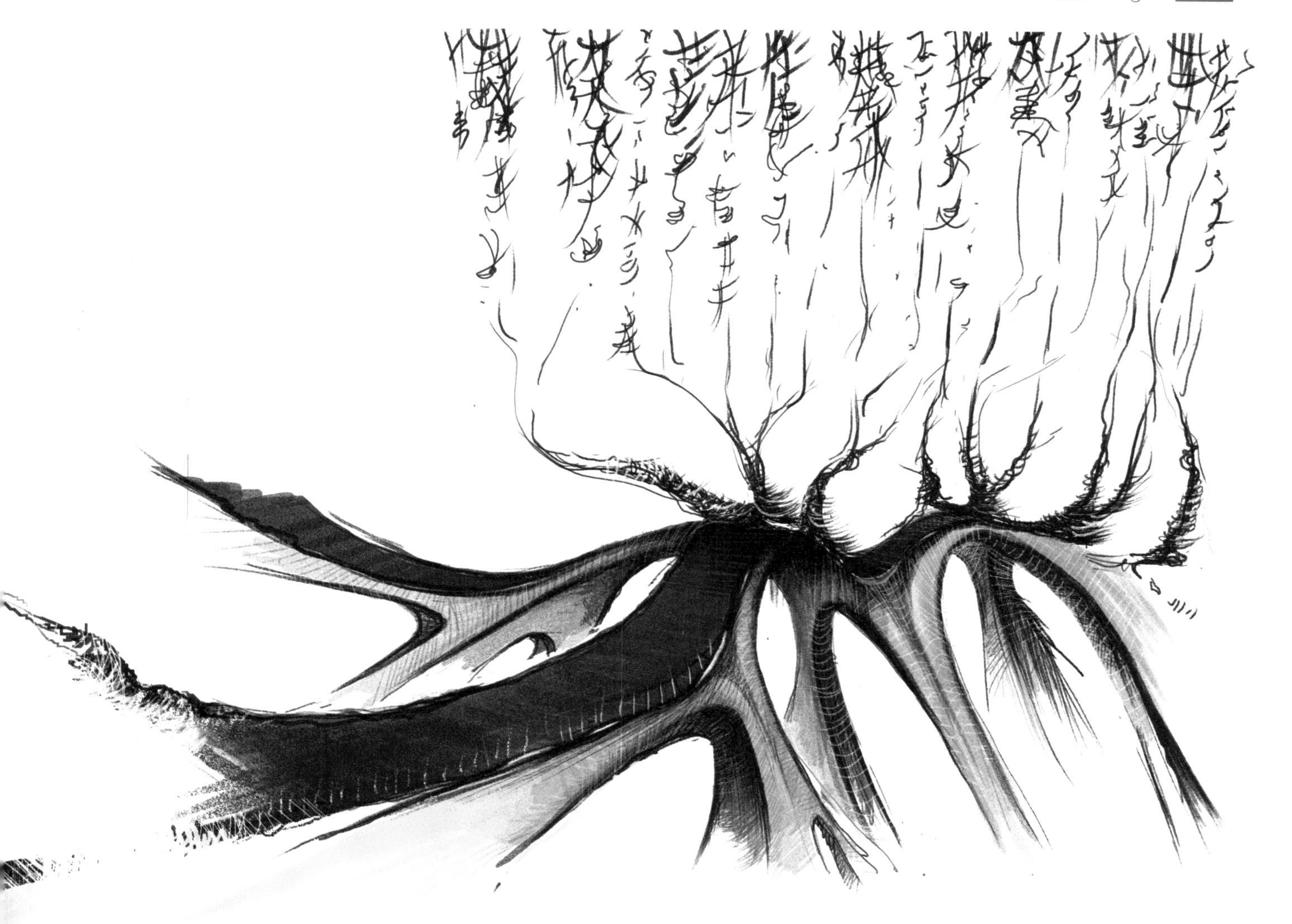

每当“造访”一些中国画时，对里面的空间印象颇为深刻，因为我没有找到经典的文艺复兴透视，而发现了对不同视点上风景的精确组织。图上的诗句为我创造了不解的故事。对我来说，他们是景观的特殊元素。当我决定采用相同的方法时，我使用了树叶，用它们来帮我写诗。

*When I "visited" some Chinese paintings I was impressed by the space, because I did not find the classical renaissance perspective, but I discovered a precise composition of scale with different points of view. The poems inside the paintings create incomprehensible stories for me. In my mind they become a particular element of the landscape. When I decided to copy the same methodology, I used the tree leaves so that they could write the poems for me.*

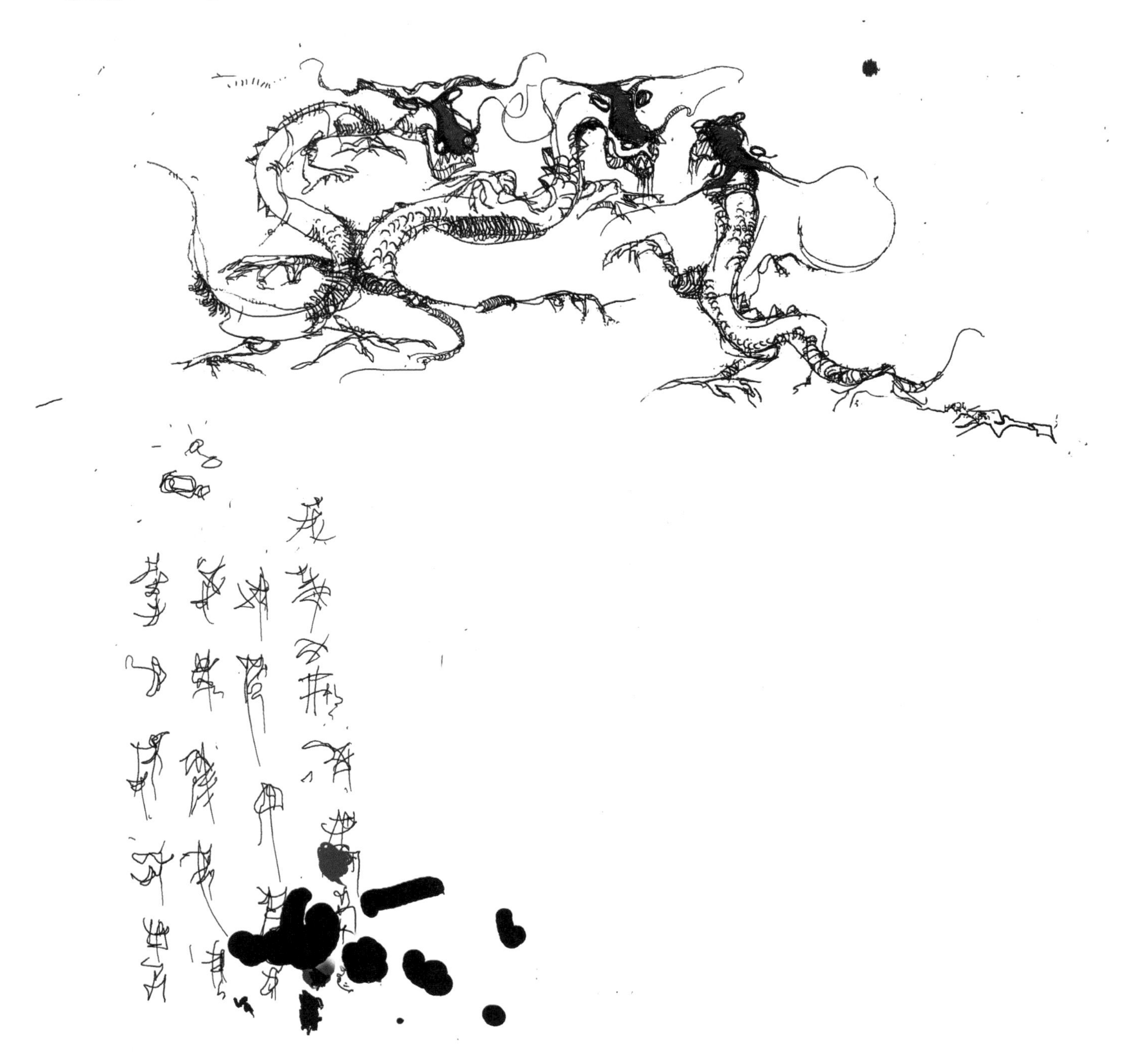

许多国家都通过绘画，标志或其他东西来象征她们的灵魂。我想中国也有代表她灵魂的东西。我不知（也没兴趣知道）龙是否可以代表中国的灵魂，但我热衷于这样认为。

*Many countries have a representation of their soul through a drawing, a logo, or something else. So I think China also has a representation of its soul. I did not know(not interested to know) if the dragons are the representation of Chinese soul, but I like to think that they are.*

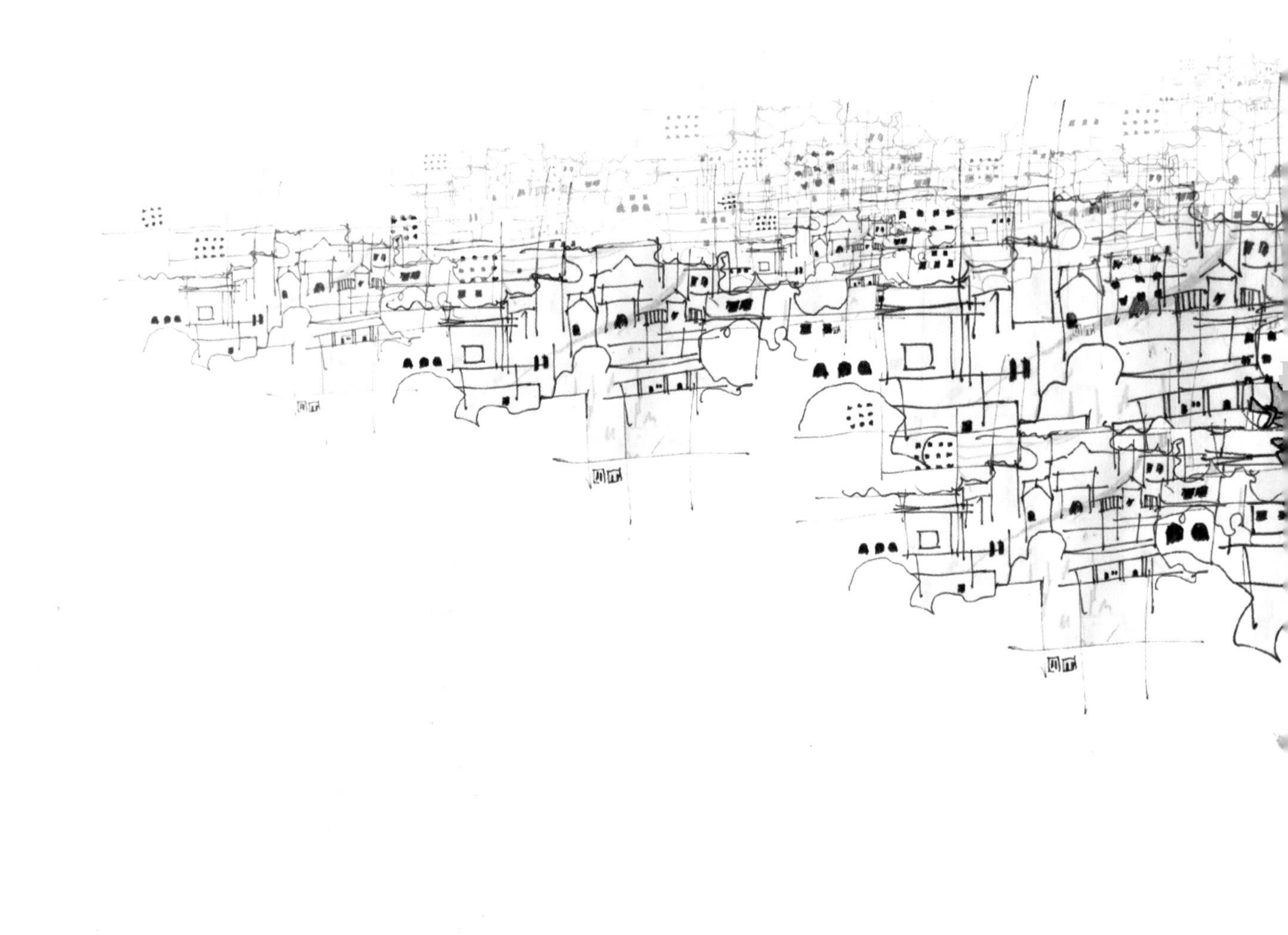

# 教学 Didactics

## 宾夕法尼亚大学景观系表现课

Representation class
Department of Landscape Architecture
University of Pennsylvania

期末评审：LARP 720“表现话题：景观绘画”，2012 年秋季

Final Review: LARP 720 "Topics in Representation: Landscape Drawing", 2012 Fall

# 教学方法
# Teaching Method

*一种明显主观的方法的成熟过程成为一种不同寻常的集体研究工具。其中每一位参与者的丰富经历促成了一种集体体验的可能性。信息的交流被延伸了，绘画变成了一种共享的体验。*

—— 瓦莱里奥·莫拉比托

*The maturing process of an apparently subjective method turned out to be an extraordinary instrument of collective research. In which the wealth of experiences of each participant has proved capable of producing a collective experience. Where the exchange of information has extended, turning drawing into shared experiences.*

—— Valerio Morabito

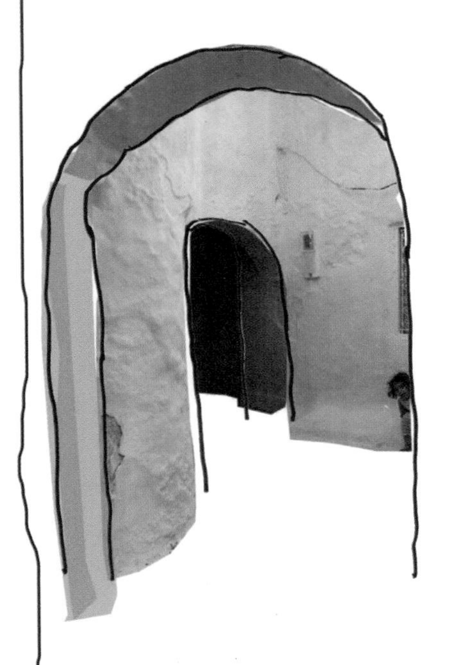

# 从画“画”到画“想法”

# From Drawing the “Drawing” to Drawing the “Idea”

*景观表现有不同程度，随着尺度和细节层面而变化；此外，它的深度和繁复与空间的复杂度相关，这反过来被统领于一个“想法”的多样异质物体塑造而成。*

——伊塔洛·卡尔维诺

*The representation of the landscape has different degrees, varying in scale and level of detail; additionally, its depth and sophistication is related to the spatial complexity which in turn is shaped by a diversity of heterogeneous objects which are linked by an "idea".*

—— Italo Calvino

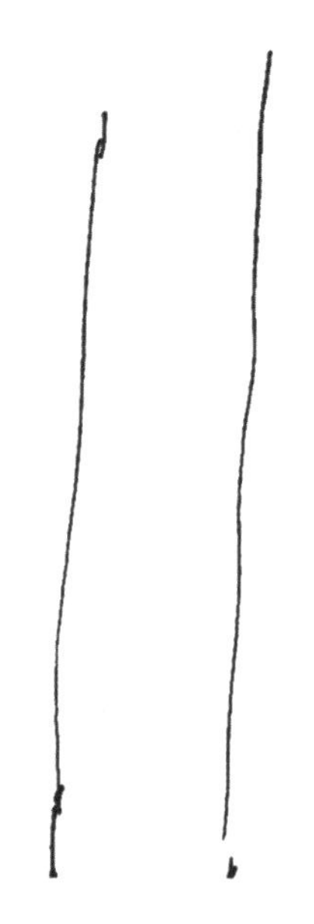

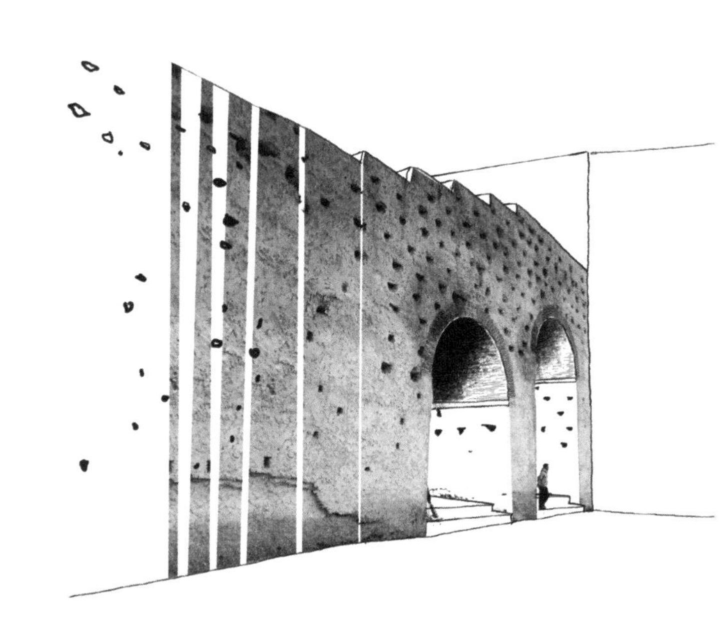

“想法”是延伸表现的起点，也是它的灵魂。对它的强调从某种程度上颠覆了传统景观表现课的理念。从画“画”到画“想法”，绘画的技巧被弱化到了较次要的位置，而创造力和想象力被提升到绝对的关注焦点。挣脱绘画技巧的束缚，学生们获得了自由同等的平台。转变观念，直到能像使用双手一样使用大脑进行绘画。跳出“就绘画而绘画”的局限，景观的延伸表现不主张花费大量的气力对景观现状进行原原本本的描绘复制，而强调利用最少的线条笔墨表达景观本质及作者的潜在意图。就景观表现的终极目标而言，延伸表现可以说是一种设计或者说是设计的开始。它带着很强的目的性，洞察现状却不描绘现状本身，发现场地改变的可能性，勾勒想象中延伸的图景。

"Idea" is the starting point for extending, and its soul. The emphasis of it, to some extent, subverts the traditional concept of the landscape presentation class. From drawing the "drawing" to drawing the "idea", the technique of drawing is relegated down to the secondary position, while the creativity and the imagination are highlighted as crucial focuses. Free from the shackles of drawing techniques, students obtain a free and equal platform. The mentality changes when the students get used to draw with their brain and not only using their hands. If the limitation of "drawing for drawing's sake" is broken, the extended representation of the landscape will not simply describe and replicate the existing landscape, which usually consumes an enormous amount of time and energy, but the emphasis will shift to the use of minimal lines, to express the essence of the landscape and the intentions of the author. In terms of the ultimate goal of landscape representation, the extended representation, in other words, is a design, or the beginning of the design process. It contains a strong purpose, while having a deep insight into the landscape, without simply depicting what it looks like; instead, it is to discover the possibility to change and to draw the bigger picture with imagination.

# 从主观性到集体性
# From Being Subjective to Collective

*它是一种产生信息，阅读和解析的练习，一种共享的过程。*
*它与寻找工具相关，当我们需要改变景观的时候，通过它可以转译，研究和理解景观。*

——瓦莱里奥·莫拉比托

*It is an exercise in the production of information, of readings and interpretations which belong to a shared process.*
*It is related to the search for instruments with which we can interpret, study and understand the landscape at times when we need to modify it.*

—— Valerio Morabito

这种本属个人的、主观的表现方法，通过瓦莱里奥在宾夕法尼亚大学景观系的教学实践，被证实为一种集体性的探索工具。他对于延伸的探索为学生带来一场头脑风暴，一步步引导他们走进景观表现的奇幻世界。拥有不同国籍、不同学科、不同知识背景的个体所生发的奇思妙想，形成独特的透镜。透过它，学生们能从同样的景观或图片中看出自己的“幻境”，并将之转为极富个性的绘画作品。正是此般形形色色，越加激发学生强烈的好奇和兴趣，创造自己的奇迹，在分享集体的奇思妙想中进一步获取灵感、激发想象。

This personal and subjective representation method, is confirmed as a collective tool of exploration through Valerio's teaching practice in the Department of Landscape Architecture at University of Pennsylvania. Valerio's exploration towards extended representation brings a brainstorm to the students, guiding them step by step into the fantasy world of landscape representation. The fancy ideas from individual students with different nationalities, discipline and knowledge background, form unique lens. Through these lens, the students see their own "wonderland" from the same landscape or picture that leads to very cool drawings and other productions. Because of such diversity, it further stimulates students' strong curiosity and interest, which further forces them to create their own miracle, obtaining further inspiration and imagination through the sharing of collective ideas.

# 发现“本质”
# Discovering the "Essence"

为了探索这些理念,课程中利用图片和谷歌地图来学习世界各地不同的场地:摩洛哥、古巴、阿根廷、智利、意大利、西班牙等。不直接亲历物质空间，通过练习来增强学生理解和提取景观品质的能力。形成一种符号字母表，学生可以像使用工具一样用它来表现我们的观点和景观的本质“思想”:“发现”景观以及改变它的可能性，就像同种疗法，能够提高场地/空间的品质而不完全改变它。这门课的目标就是提供学生一种表现工具以开始这一过程：即如何表现本质或者极少的思想，但如果只是绘画景观现状或者处于刚刚转变的起点，这个阶段是不明显的。(引自 Larp 720 课程大纲，2011 年秋季，“如何发现项目表现景观”)

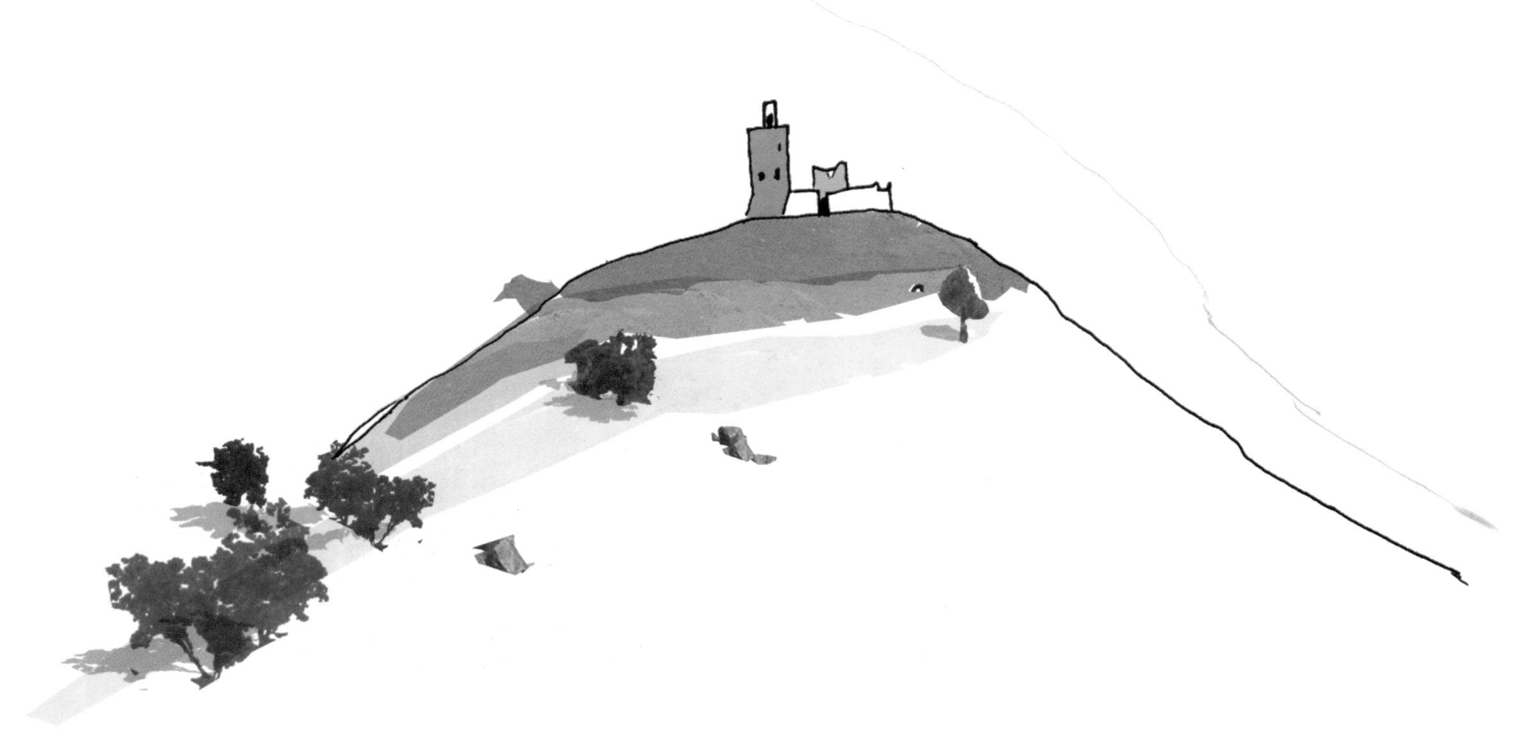

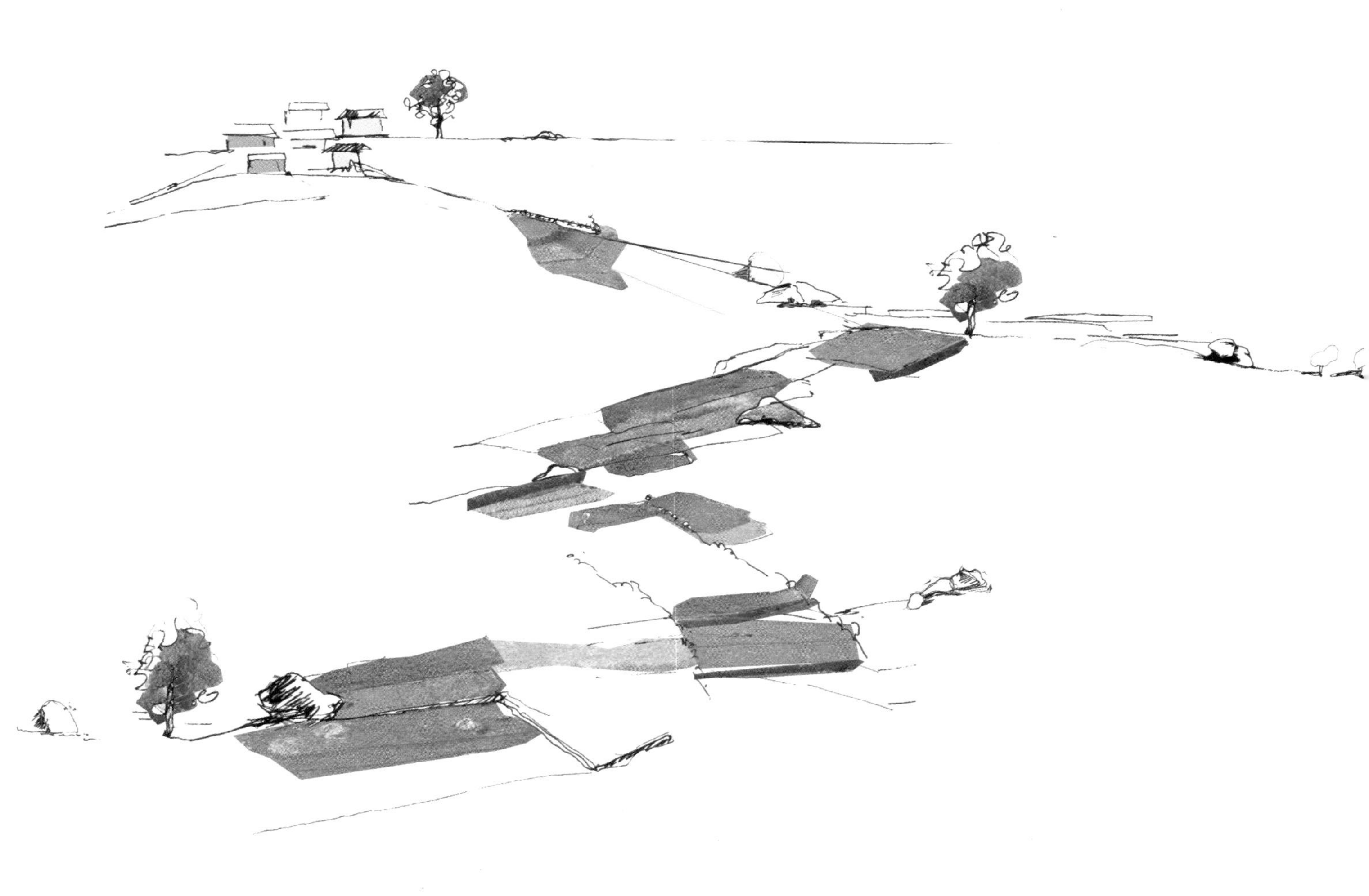

In order to explore these notions, we will study sites in different part of the world with pictures and Google earth: Morocco, Cuba, Argentina, Chile, Italy, Spain and so on. Doing exercises to improve our capacity of understanding and capturing the quality of landscape without direct physical experiences. A way to form a kind of alphabet of sings we can use like tools to represent our point of view and the essential "idea" of our landscape: the "discovering" of landscape and possibility to change it, in a similar way as homeopathic medicine is able to improve the quality of the site/space without changing it completely. The objective of this course is to provide students with the representation tools to begin this process: how to represent the essential or minimal idea, a stage in which it is not, evident if we are drawing the existing landscape or the beginning of its transformation. (From Larp 720 Course Syllabus, Fall 2011, "How to discover a project drawing a representation of a landscape" )

# 发现“可变性”

# Discovering the "Changing"

瓦莱里奥认为景观是一个复杂系统，是众多要素相互关联的结果，这界定了复杂空间平等的条件。在景观表现和项目之间存在一种关系，它能够引发改变自身的过程。清晰的符号或是隐含的迹象与变化的过程存在交互关系。恰当地描绘表现景观的过程与获取知识的过程密切相关，所以有必要抓住场地的本质，它的文化、物质特征和形态。观察场地时，发现景观的品质或者“风格”是可能的。同时，也可能发现景观的“可变性”，这种“可变性”是延伸的大图景的关键成分；要素，碎片及其他特殊的方面都是透露场地新空间的基础。路边的树、特别的农田、景观中的建筑、矿坑、电线杆等等都是被用来探索和确定设计最初想法的要素。（引自 Larp 720 课程大纲，2011 年秋季，“如何发现项目表现景观”）

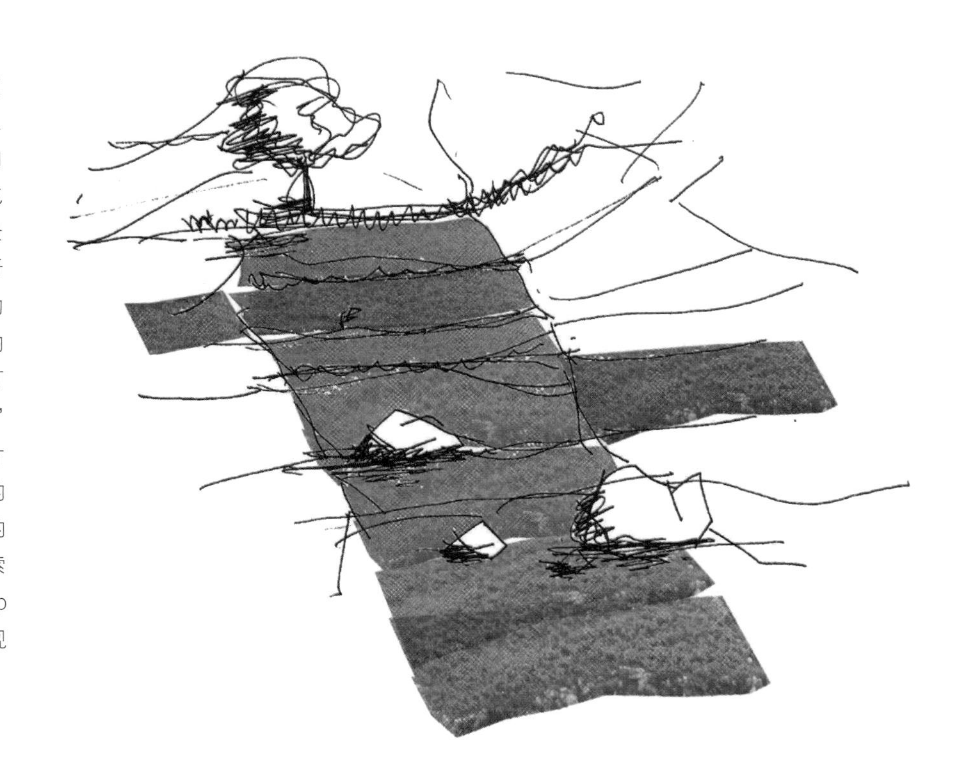

Landscape is a result of a complex system of interrelated elements, which define equality complex spatial conditions. There is a relation between the representation of a landscape and the project, capable of igniting the process that can change it. Clear signs or hidden traces are tied to the process of change in a reciprocal relation. The process to appropriately portray the representation of a landscape is tied to the process of knowledge; thus, there is a need to capture the essence of a site, its culture, its physical characteristic, its morphology. While observing a site, it is possible to discover the quality or the "style" of the landscape. At the same time, it is possible to discover the "changing" of the landscape, the essential components of the bigger picture; the elements, fragments, particular aspects which are basic to reveal the new space of the site. Trees along a road, particular agricultural fields, buildings in the landscape, quarries, electric lines and so on are elements to be used to explore and fix the first idea of the design." (From Larp 720 Course Syllabus, Fall 2011, "How to discover a project drawing a representation of a landscape" )

# 领会与想象
# Seeing and Imagining

从绘画“想法”到描绘“想象”，表述了瓦莱里奥延伸表现课的进展过程，而想象力的培养是课程的关键目标。这门课不仅为学生提供了一种独特的表现工具，更重要的是引导他们开启自己的想象之门，挖掘这种取之不尽的脑力资源。本着这一目标，课程分为两个阶段：“上半部分：研究景观中的‘想法’并学着表现它。学生们利用传统速写、电脑绘画、抽象模型、图片剪辑，以及图片与速写之间关系，来发现和理解景观中基于个人的和集体的‘想法’。下半部分：制作幻境，极具创造性和发明性，与个人的思想和知识有很强的关联度。渴望发现出奇的想法以及非凡的意义，任务包括速写与绘画、电脑绘画、概念模型、速写与图片构成、集体画作。”（引自Larp 720课程大纲,2007年秋季,“景观的领会与想象”）

Shifting from drawing the "idea" to drawing the "imagination" is a progress of Valerio's representation class, while the cultivation of imagination is the key objective of the course. This course provides the students with a unique tool for representation, and what is most important is to guide students to enter the realm of imagination and to dig the abundant intellectual resources. In line with this goal, the class is divided into two phases: "The first half: research the 'idea' of landscape and learn to represent it. Students used traditional sketches, digital sketches, abstract models, alternation of pictures and the relationship between pictures and sketches, to discover and to understand personal and collective 'ideas' of landscape. The second half: product of fantasy, the one of creativity and invention, resulting from the relations (connections) between one's thoughts and knowledge. Aspired to discover new points of view and unexpected meanings, work ranged from sketching and drawing, computer sketching, study models, composition of sketches and images, to collective painting. "(From Larp 720 Course Syllabus, 2007 Fall, " Seeing and Imagine Landscape")

# 学生作业
Students' Works

*在这一过程中，徒手画被证明远远不止是一种主观的工具：在数字技术中，它成为景观延伸表现的必要工具，这种表现超越了图像表面的东西。*

——瓦莱里奥·莫拉比托

*In this process the freehand sketch has proved to be more than just a subjective instrument: within digital technology. It has become a necessary instrument for the extension of the representation of landscape beyond the appearance of images.*

—— Valerio Morabito

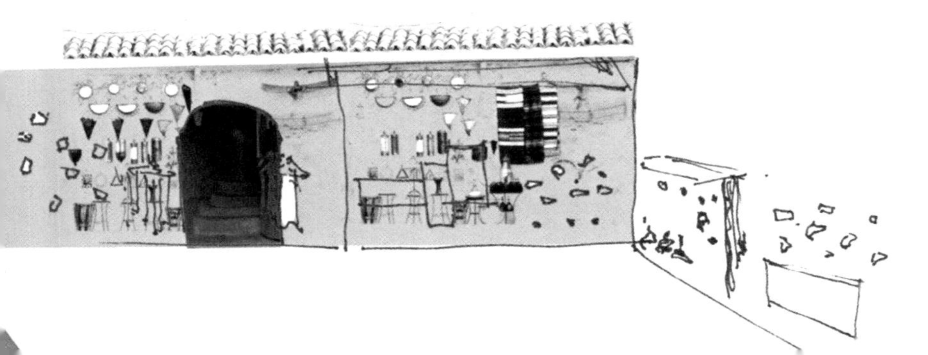

# 记忆与表现

*不要画你所见*

*不要怕画错，错误是产生变化的可能*

*有时记忆比图片更强大*

—— 瓦莱里奥·莫拉比托

记忆的空白之处是延伸表现的延伸之处，这是瓦莱里奥表现方法的最基本观念。他不主张看到什么就画什么，他的创作往往都基于旅行或者阅读之后的记忆进行。记忆，不准确，却像个筛子。剩下的片段也许正反映了事物最特色最本质的方面，而筛去的东西及其留下的空白恰恰提供了创造的机会。而在传统认识层面上，记忆的空白处是可能导致错误的地方，一旦记忆模糊就易引起紧张心理。若能转变这一观念，记忆的负担反能成为创作的源泉。根据记忆的不同来源，瓦莱里奥将他日积月累的作品，分别装入“景观中的记忆”、“城市中的记忆”、“记忆中的记忆”这三大“口袋”。教学的过程也是训练学生如何一步步转变观念，基于记忆进行绘画的过程。他也交给学生这三个“口袋”，随着教学的日渐深入让其“口袋”里的东西越来越丰富。

*Don't draw what you saw*
*Draw the mistake, make possibility to change*
*Memory is sometime more powerful than pictures*

—— Valerio Morabito

# Memory and Representation

The blank area of memory is the place where the extension may happen, which is the fundamental idea of Valerio's representation method. He does not advocate drawing exactly what you see; rather, he draws on the basis of the memory, which comes from travels or readings. Memory, though it is not accurate, serves like a sieve. The remaining things may reflect the most special and essential aspects, and when some things are gone, the left space may provide the opportunity to create. Nevertheless, in a traditional way, it is common to think that the blank area of the memory may lead to mistakes, and the fuzziness of memory generates the worries of making mistakes. Once this idea is changed, the burden of memory becomes the source of creation. According to the different sources of memory, Valerio accumulates his works into three categories, his three "pockets": "Memory in the Landscape", "Memory in the City" and "Memory in the Memory". The process of teaching is a process of training the students step by step to change their ideas, to draw on the basis of memory. He also provides the students with the three "pockets", and the furtherer the class progresses, the more abundant the "pockets" get.

# 景观中的记忆
# Memory in the Landscape

*画风……不要画树*
*画褶皱……不要画石头*
*画光……不要画灯*
—— 瓦莱里奥·莫拉比托

*Draw the wind……not the tree*
*Draw the winkle……not the stone*
*Draw the lignt……not the lamp*
—— Valerio Morabito

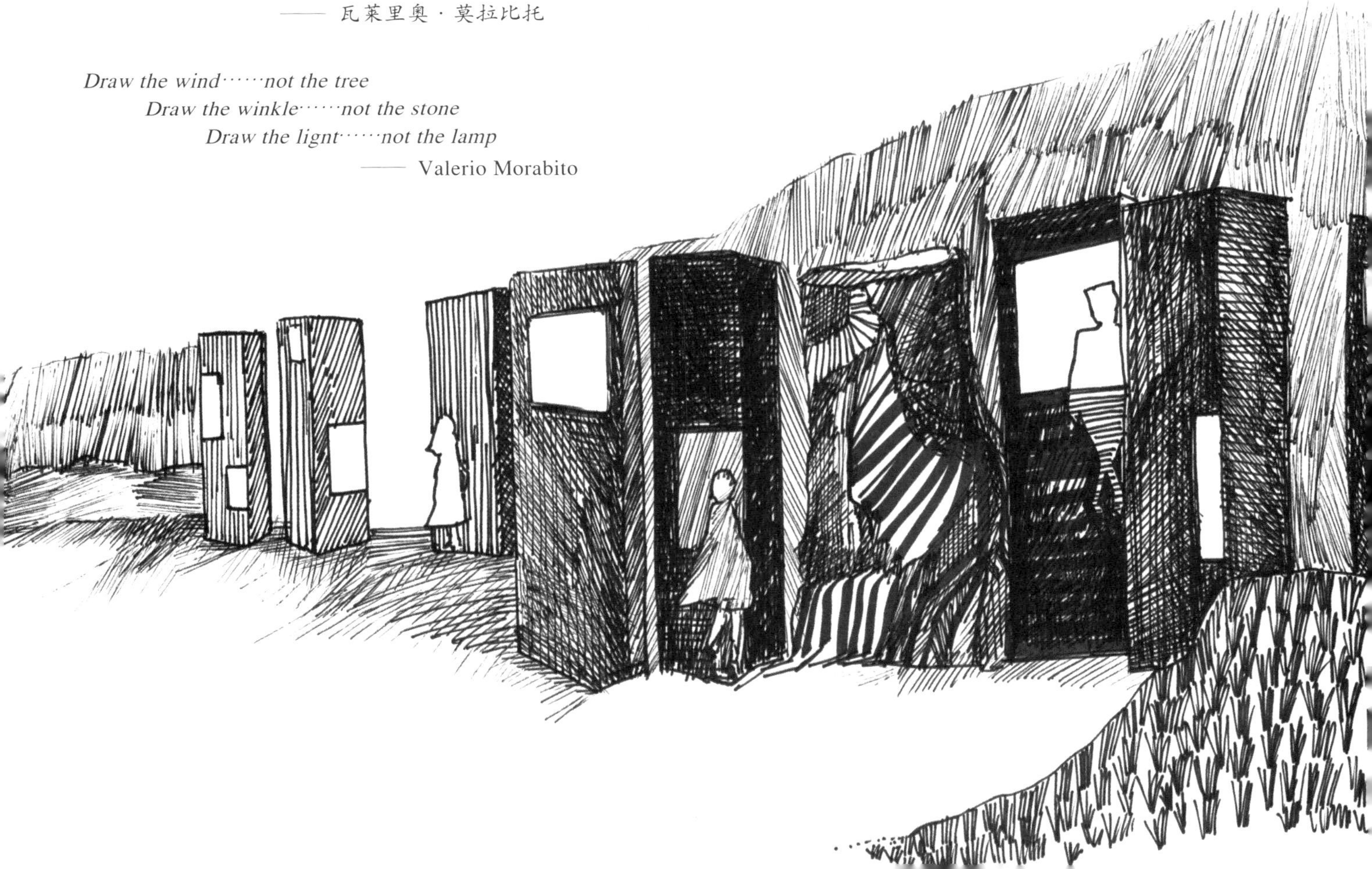

速写训练是上半学期课程的重点，半堂课练习，半堂课挂墙点评。貌似寻常的速写课却是未曾见过的惊心动魄，分分秒秒牵动着学生的心神。形象一点，可以把瓦莱里奥的方法称作“2 秒钟训练法”。搜罗世界各地最具特色的景观与城市，每堂课上瓦莱里奥会展示一处，将他精心选择的图片作为学生的速写素材。对选定的地点进行简要的介绍之后，他开始播放图片，几乎每张图片都只展示 2 秒钟。随机抽中一张，停留 2 秒，关闭图片，让学生凭记忆以最快的速度进行速写。在速写的过程中，瓦莱里奥在教室来回走动观察学生如何绘画，一边看一边以幽默的方式传授他的理念方法，将学生一点一点从传统的表现方式和习惯转到对延伸表现的探索。“2 秒钟训练法”的意图，在于让学生学会过滤复杂的图片信息，达到“最少”表现，学会利用最少的线条提炼景观本质，对线条有精准的把握。

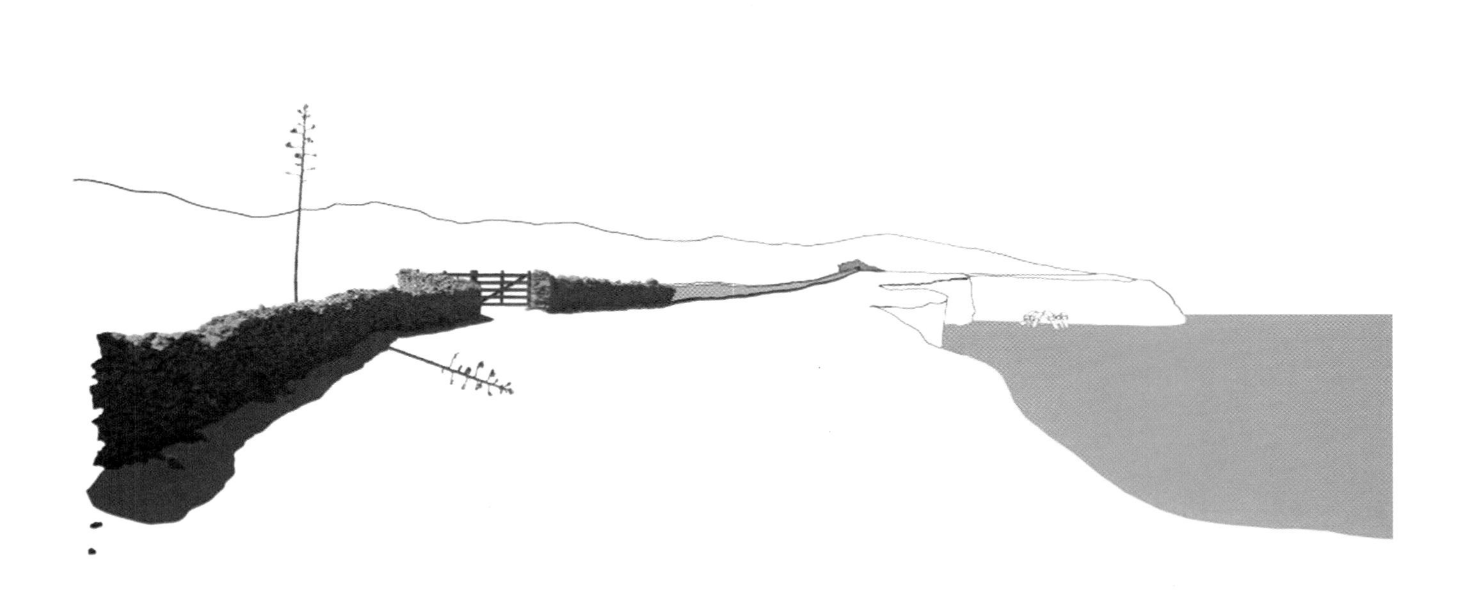

Sketch training is the main task of the first half semester, with half of the time spent on sketching, and the other half spent on pin-up critique. It seems to be an ordinary drawing class, but it is extremely astonishing for it attracts students' attention every second. To describe it in a vivid way, Valerio's training method can be called "Two-Second Training Method". He searches around the whole world for the most distinctive landscape and city, and he always chooses one of those places for each class, picking up elaborate pictures that he uses as teaching material for sketch training. After a brief introduction to the selected place, he begins to display pictures, 2 seconds for each one. He randomly picks one, he keeps it on display for 2 seconds, and he then closes the picture and asks the students to draw as fast as possible according to their memory. In the process of sketch, Valerio walks around to observe how students draw, and at the same time he explains his philosophy and method in a humorous way, step by step, guiding the students to transit from traditional ways and habits of drawing to the exploration of the extended representation. "Two-Second Training Method" helps to teach the students to filter the complex information of pictures and to get to the "least" expression, so that the students can use the minimal lines to extract the essence of the landscape and get a precise grasp of the lines.

虽然展示的是图片，瓦莱里奥要求学生画的却是超越于图片表象的东西。引导学生将自己置身于画面的环境中，走进去感受，去发现里面独特的东西或者更深层更本质的东西，因为这些是决定景观独特性的基因。这一过程让学生学会如何结合自己拥有的记忆和知识去消化图片，通过消化看到非常个性化的景观，并寻找无奇不有的方法与材料去创造、去表达。

“景观”和“城市”的速写训练穿插着进行，从对自然环境、自然要素及自然力的延伸表达到对城市环境、建筑空间及文化情结的特殊理解，延伸表现思想潜移默化地改变着学生们的思维和表现习惯，并构建全面的新颖的表现语汇。对于景观记忆的训练，瓦莱里奥的关注点在于景观的成因，如风、水、光等自然力对树木、石头等要素的形态塑造；云、天空、火山、地平线之间的对话；地形、土壤、植被、构筑物之间的生成关系……慢慢的，学生们开始习惯于绘画前对景观的解读和消化，关注暗藏于景观内部的能量和关系，尝试以自己的方式去表现。课后瓦莱里奥要求学生根据每堂课所画的速写制作概念模型，并将速写、图片、模型等成果进行版面设计，寻找之间的联系，创造一种秩序或一个故事，到下堂课进行挂墙点评，集体分享。

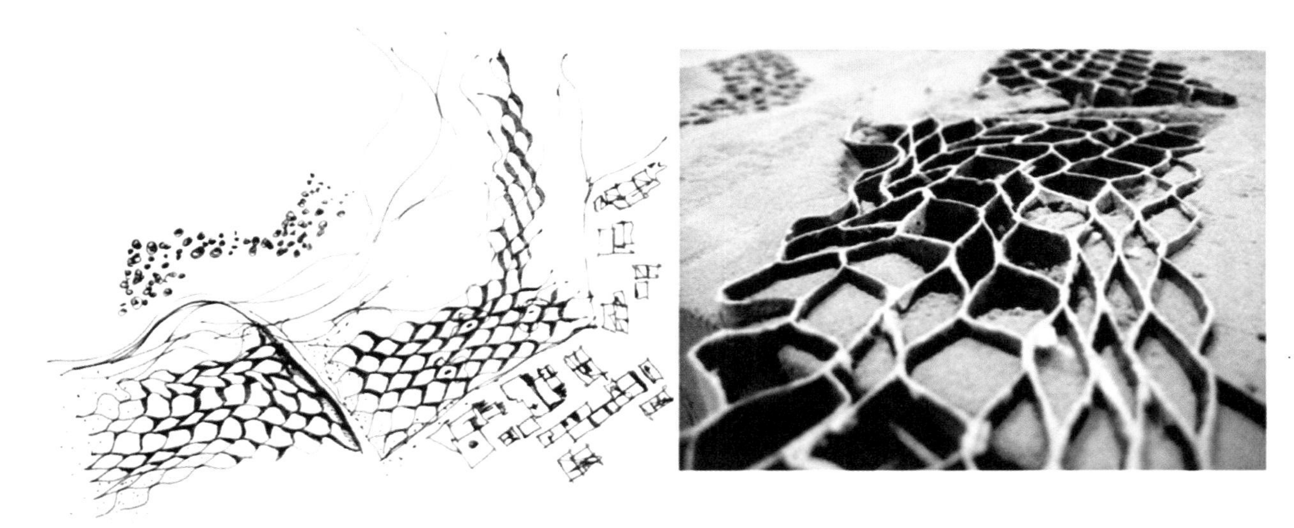

Though the pictures are displayed, Valerio requests something that goes beyond the image itself. He guides the students to get inside the scenario of the picture so to feel and discover the things which are special, deep and essential, because such things are the genes which define the uniqueness of landscape. This process leads the students to learn how to combine their own memory and knowledge to digest pictures, so that they can generate their personalized idea about landscape and find out the ways and the materials to create and express.

Sketch training regarding "Natural Landscape" and "City" is done in turns: from the extended expression of natural environment, natural elements and natural forces, to the special interpretation of urban environment, building space and cultural emotion. The idea of extended representation unconsciously changes students' thinking and habits on drawing, and builds a comprehensive new vocabulary. For the training regarding the natural landscape, Valerio focuses on the causes of the landscape: how the natural forces such as wind, water and light shape the trees, the rocks and the other elements; the dialogue among the cloud, the sky, the volcanoes and the horizontal lines; the relationship between the topography, the soil, the vegetation and the structure, etc. Gradually, the students begin to get used to reading and digesting the landscape before drawing, and they start paying attention to the interior energy and relationship which is not evident, and they explore their own ways to represent it. After the class, Valerio asks the students to make a conceptual model according to the sketches, and then makes a panel to build the link between the sketch, the picture, the model and so on, to create a sequence or to invent a story, which will be pinned up to a board so they can be critiqued and shared collectively.

画风

Draw the Wind

4. Root into the rocks
5. Trees of same height
6.layers of mountains
7. Forest  and the mountain
8. Wind form the trees

画能量

Draw the Energy

画本质

Draw the Essence

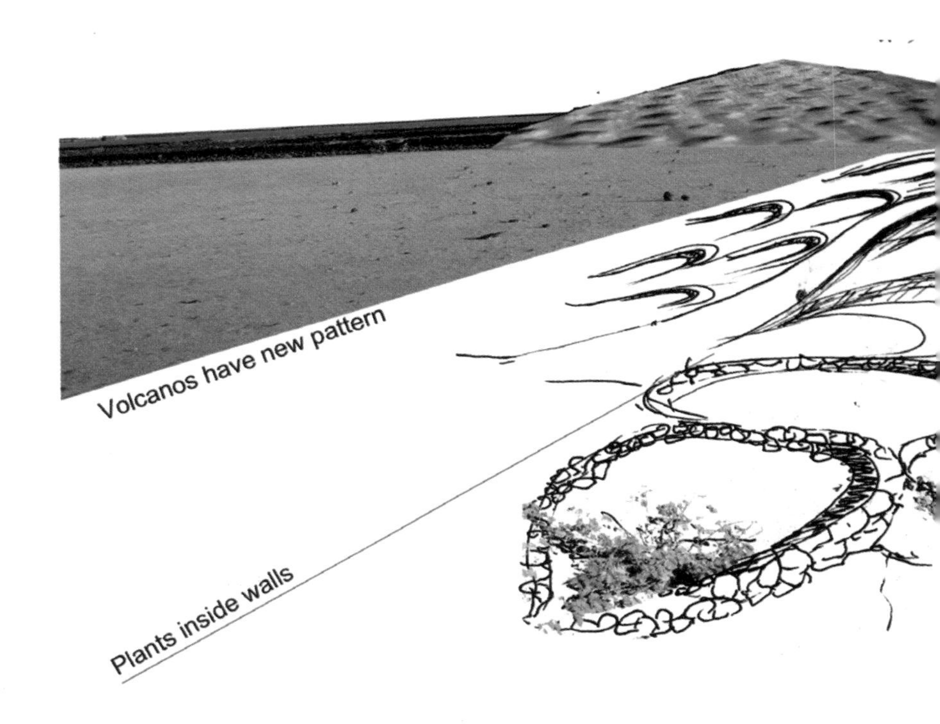

Rocks creep on sand

# 编故事
Invent a Story

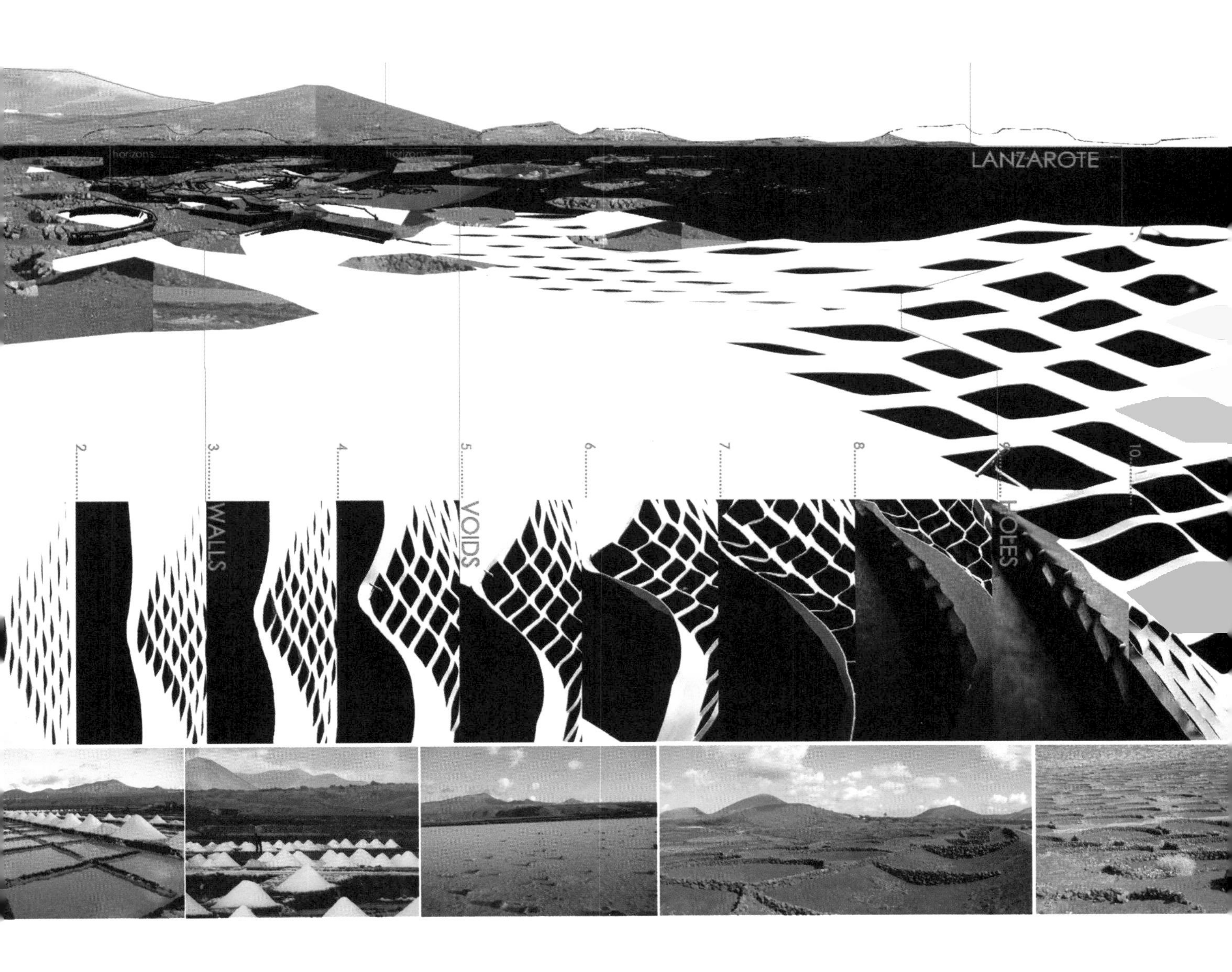
horizons
horizons
LANZAROTE
2
3
4
5
6
7
8
9
10
WALLS
VOIDS
HOLES

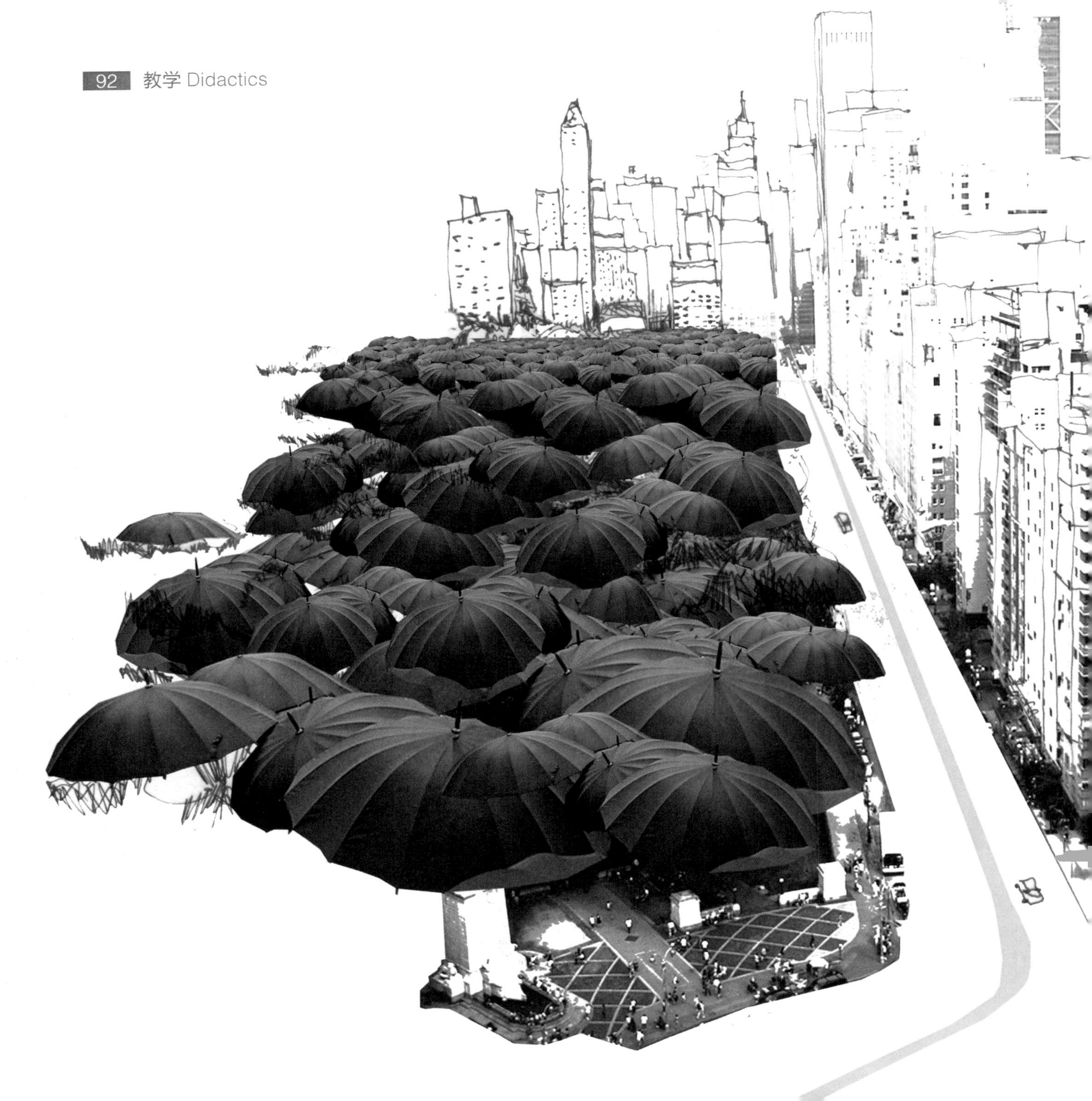

# 城市中的记忆
# Memory in the City

*走进图片*
*想象你在空间里面*
*置入新的东西*
*添加你的经历*
*创造故事*
*寻找事物之间的关系和结构*
*不要表现现状本身*
*用最少的线条画出最多的信息*

——瓦莱里奥 · 莫拉比托

*Go inside the picture*
*Imagine you are inside*
*Layering new things*
*Invent stories*
*Add your experiences*
*Find relationship or structure between something*
*Do not just to represent the existing things*
*Use the least line to draw the most information*

—— Valerio Morabito

对于城市记忆的表现，瓦莱里奥强调不同视角、不同尺度、不同文化因素影响下的空间品质及空间关系。但不论哪方面，融入个性的、趣味的想法，以及如何引发对更大空间图景的联想是他对学生作品的最基本要求和评价标准。简而言之，就是用绘画来表达自己对城市的感想与观点，添加自己的特殊城市体验，并融入对城市新的期望，基于现状创造独特的城市幻境与故事。类似于很多电影对城市空间的描绘，从高空俯瞰的全景到城市内部的节点空间序列，对城市的延伸表现也要求这种多视角多维度的蒙太奇式的空间解读和表现，为发现空间品质和延伸潜质提供了可能性。城市终究是为生活的，丰富的生活场景是空间品质的关键因素，它赋予城市特殊的性格，其中的色彩、音乐、舞蹈、活动等文化元素是延伸表现的主要“想法”来源。

As for the representation of the memory in the city, Valerio emphasizes the spatial quality and spatial relationship under the influence of factors such as different perspectives, different scales and different cultures. It does not matter which aspect is considered, adding in personalized and interesting ideas, and knowing how to trigger the imagination of a bigger picture, are the most basic requirements and evaluation standards for the students' work. In other words, it is a matter of drawing personal impressions and viewpoints towards the city, adding in a special experience, putting a new expectation into the city, and creating a unique urban dreamland and stories based on the existing conditions. Similar to the way the urban space is depicted in several movies, from the panoramic view from the sky to the nodal space sequence inside the city, the extended representation of the city also requires this montage-style space interpretation and representation method with multi-perspectives and multi-scale dimensions. It provides the possibility of discovering the space quality and the potential to be extended. After all, the city is for the life, the diverse scenes of life are the key factor of space quality, which gives the city its special character, and the cultural elements such as color, music, dance, and activity are the main sources of the "idea " of extended representation.

城市与树木

City and Tree

塔与森林

Tower and Forest

The town of fine towers

## Saint Gimigniano

Tuscany, Italy

Forest between two towers

Buildings in the Medieval town

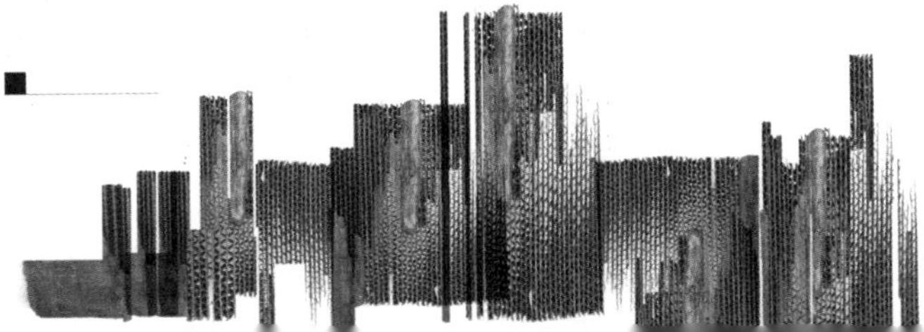

MORRIGAN
APHRODITE
PERSEPHONE
CERIDWEN
ERESHKIGAL
HATHOR
ARIANRHOD

城市与色彩

City and Color

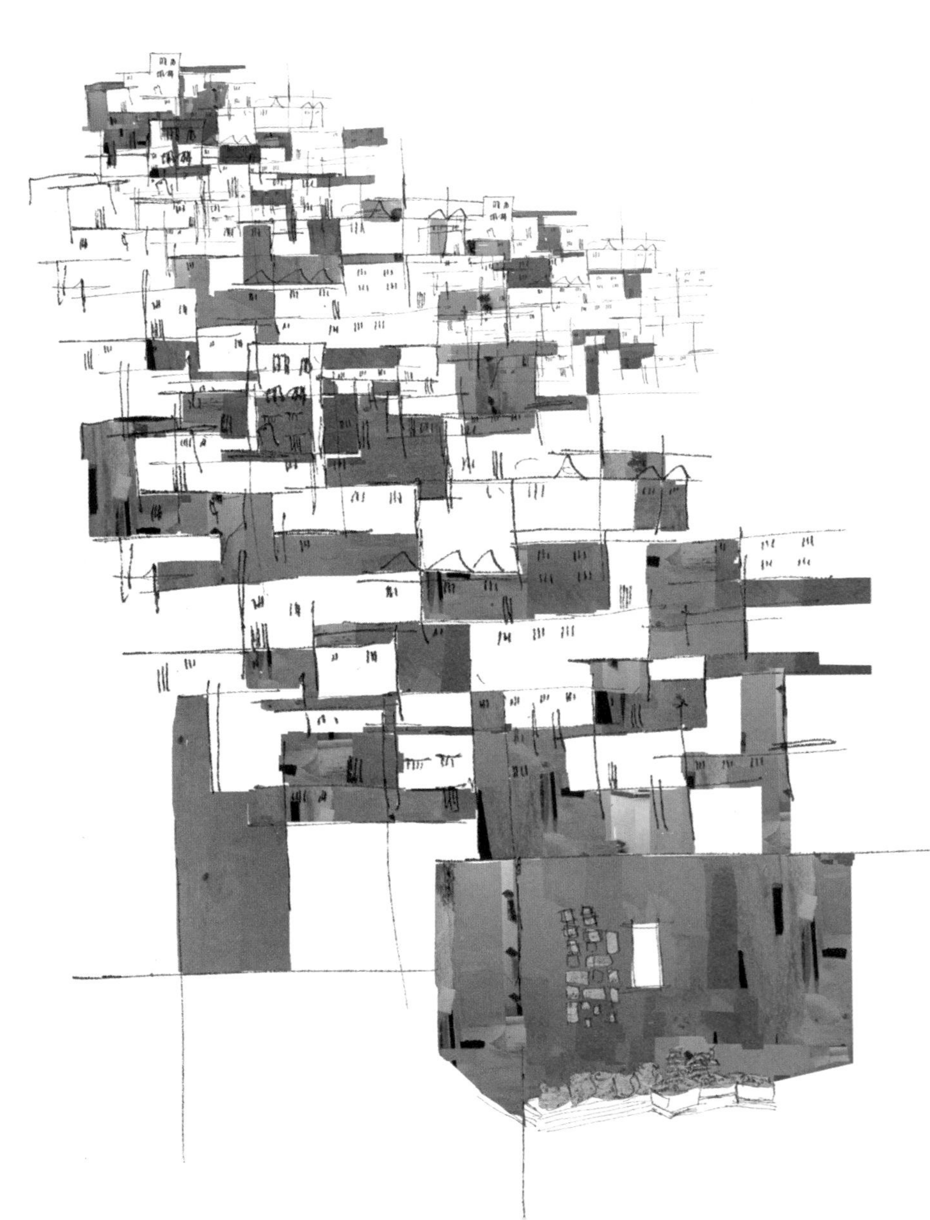

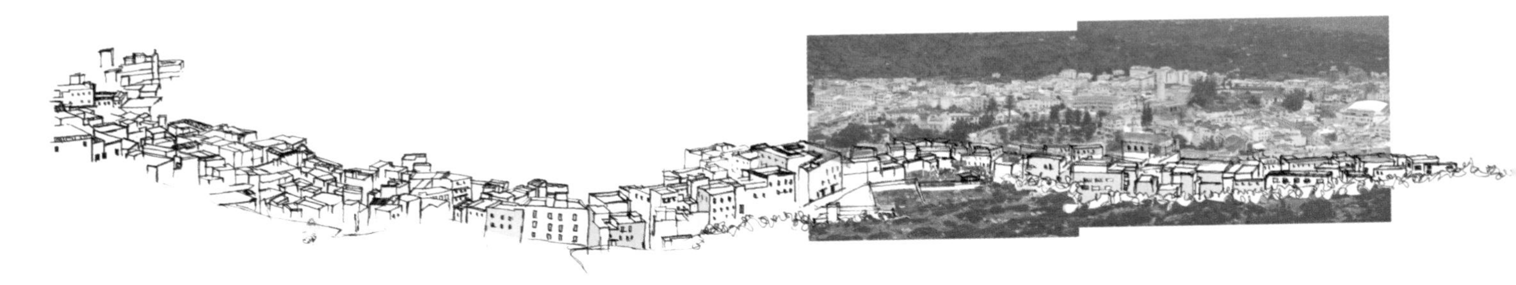

城市与故事

City and Story

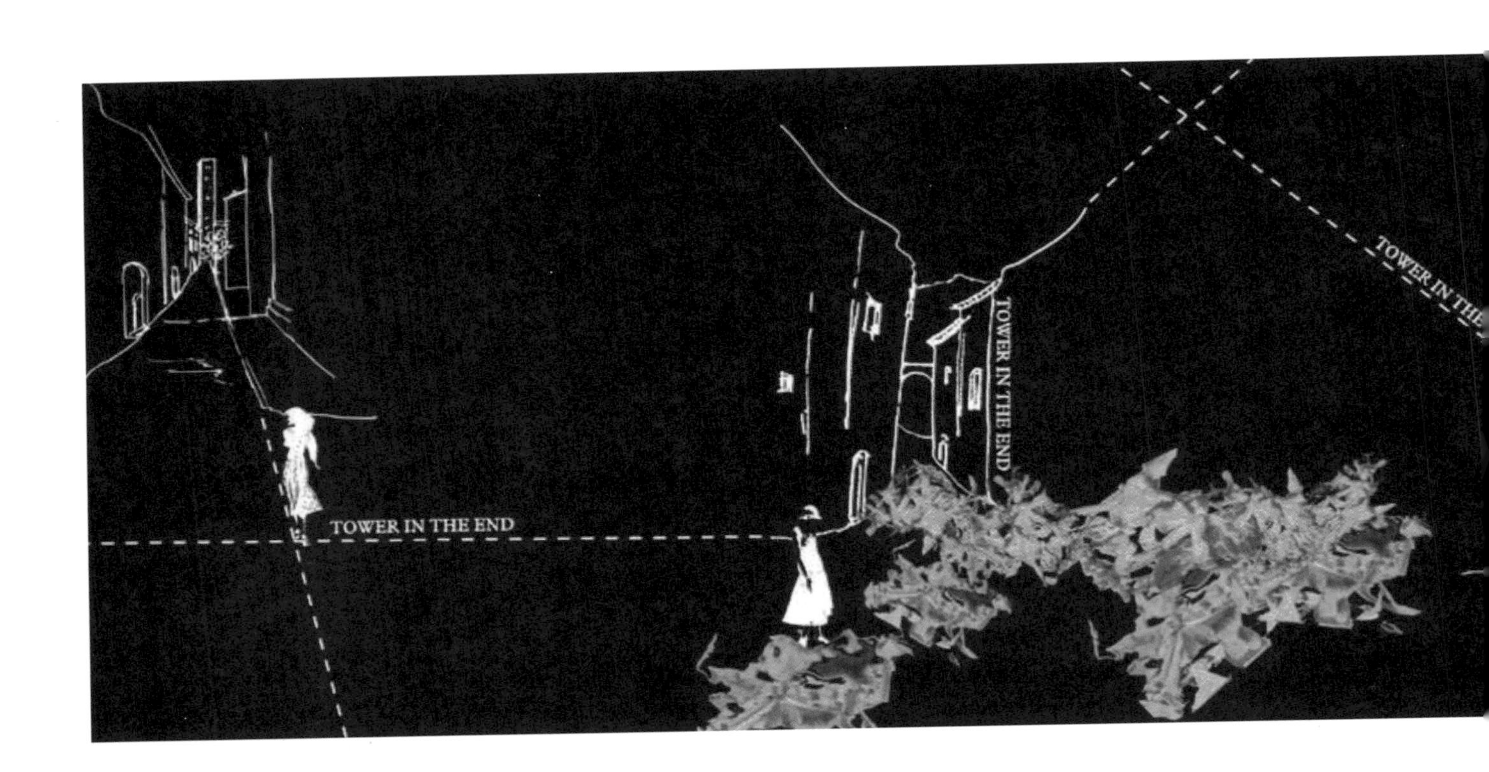

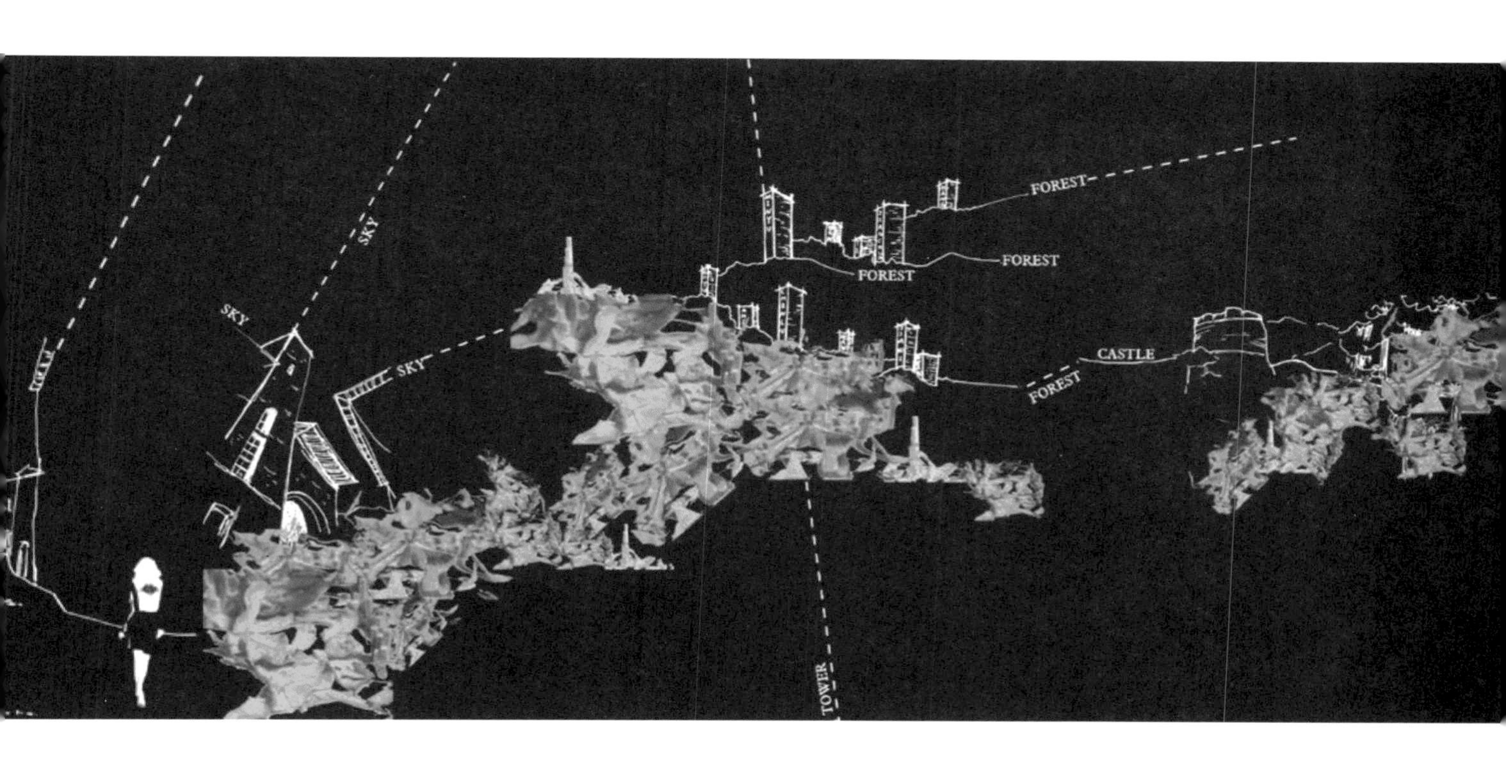
SKY
SKY
SKY
FOREST
FOREST
FOREST
FOREST
CASTLE
TOWER

Vertical City
Towers
Houses
Layering of Architecture
Layering of Nature
Horizontal / Extending
Vertical /
Narrowing
Suburban
Mix/
Mess
Urban
Restraint
Wild

every time you enter the square,
you find yourself caught in a dialogue

As time passes the roles, too, are no longer exactly the same as before

the participants in the dialogues die one by one

and meanwhile those whi will take their places are born

# 记忆中的记忆

# Memory in the Memory

*想象力是人类最宝贵的财富，不需任何花费，却能给你带来无穷的益处，因此一定要竭力挖掘它。*

——瓦莱里奥·莫拉比托

*Make the best use of your imagination, this is the richest, powerful and infinite thing you have to consume.*

——Valerio Morabito

*想象是落入东西的地方。*

*对我们来说，今天的城市是什么？我认为我写了一种东西，它就像是在越来越难以把城市当做城市来生活的时刻，献给城市的最后一首爱情诗。*

——伊塔洛·卡尔维诺

*Imagination is the place where falls into something.*
*For us, what is the city today? I think I wrote one thing, it likes the last love poem I consecrated to the city, at the time it gets harder and harder to use city as a city for life.*

——Italo Calvino

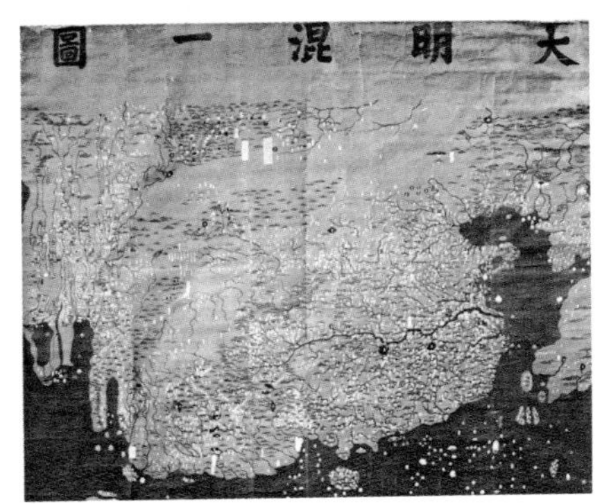

大明混一图
Amalgamated Map of the Great Ming Empire

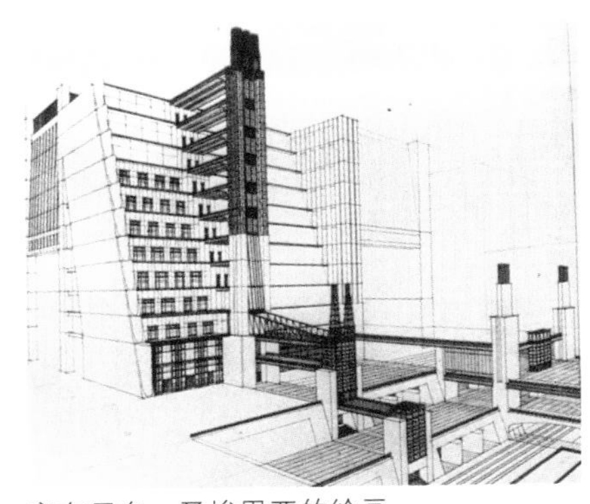
安东尼奥・圣埃里亚的绘画
Drawing of Antonio Sant'Elia

*italo* calvino

文学资料是瓦莱里奥最主要的灵感来源，最具激发想象的能量。下半学期的任务是创造一座想象中的城市，而卡尔维诺的《看不见的城市》成为最重要的教学材料。这是一本短篇小说，作者借马可波罗之口塑造了一个个虚构的城市，超越于空间和时间的想象的城市。卡尔维诺用语言带领读者体验他脑海中奇妙的城市图景。瓦莱里奥期望学生借鉴卡尔维诺构想并描述城市的方式，包括城市空间、故事场景、记忆符号等，发挥无边想象，创造自己的梦幻城市，并用各种方式手段和材料对城市的平面、剖面、效果、模型及展板加以表现。基于现实又超脱现实，从抽象的脑海中的城市图景到富有细节有空间感的图纸模型，两者之间的跨越困难重重。

困难，却赋予课堂意想不到的丰富。瓦莱里奥设计了许多有意思的“小插曲”为学生们排忧解难。首先是电影展示，与城市空间相关的几部电影片段，《大都会》、《柏林苍穹下》、《里斯本故事》、《德州巴黎》、《银翼杀手》，被用来强调不同角度下观看到的城市实体空间，以及城市光线、色彩、音乐、舞蹈、故事等无形却对城市氛围、文化起关键作用的要素。接着是乌托邦建筑师安东尼奥・圣埃里亚的建筑与城市绘画的展示，使学生对乌托邦城市空间的特质和表现方法有了感性认识。此外，各种有趣的中外古地图也被搜罗来指导学生进行平面图的表达，因为古地图画面原始直观、信息丰富、表达准确、极富空间感，与瓦莱里奥所期望的表达方式不谋而合。还有一个最重要的元素是中国国画。瓦莱里奥对之再三强调，并要求学生进行深入研习。尤其是国画中的留白处理，画面层次，以及空间序列的组织等画理是瓦莱里奥所赞赏的。他甚至说他的延伸表现方法也具有中国国画的诗意，在本质上与中国画有很多相通之处。

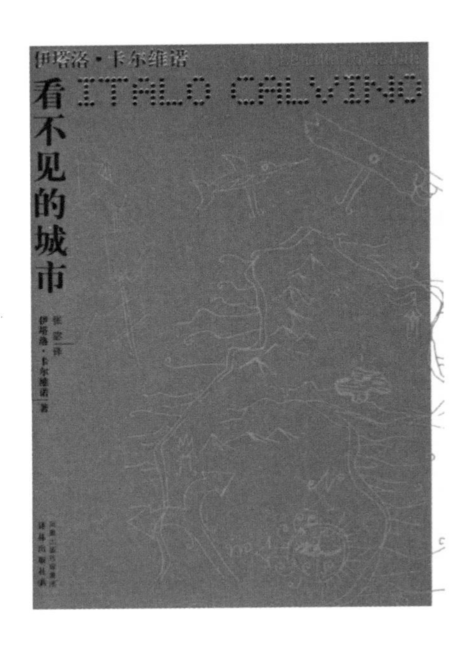

Literature, with great power to ignite imagination, is Valerio's main source of inspiration. The task of the second half of the semester is to create a fantasy city, and Calvino's *Invisible City* is used as the most important teaching material. In this short novel, the author shaped many fantasy cities through Marco Polo's narration, the imaginary cities beyond time and space. Through language, Calvino led readers to follow his mind and experience the wonderful cities. Borrowing Calvino's way of conceiving and describing the city, including urban space, stories and scenario, memories and signs etc., Valerio expects the students to create their own imaginary city with boundless imagination, and expects them to employ various ways and materials to represent it, in terms of master plan, sections, rendering, models and panels. The reality is taken as a basis, and yet reality is surpassed; It is extremely difficult to bridge the gap between the abstract mental image and the specific drawings and models.

Nevertheless, the difficulties endow the class with an unexpected richness. Valerio organizes a series of interesting "episodes" to help the students get through. First, the movies. Several movie clips related to urban space, such as "Metropolis", "Wings of Desire", "Lisbon story", "Paris, Texas" and "Blade Runner", are used to highlight the urban physical space from different perspectives, as well as the vital role of the intangible elements such as light, color, music, dance, and stories, to form the urban and cultural atmosphere. Then come the drawings about architecture and city from the utopist architect Antonio Sant'Elia, which help the students to have a perceptual understanding about the characteristics of the utopian urban space and its representational method. Besides, all kinds of interesting Chinese and foreign ancient maps are also picked up to guide the student into drawing the master plan, because the ancient map seems primitive, intuitive, informative, precise and spatial, which is the way Valerio expects the representation to be. The most important element is the Chinese traditional paintings. It is strongly recommended by Valerio for the students to have an in-depth study. Especially the drawing theory, such as the white space arrangement, the levels of the drawing, the organization of the spatial sequence, is of great interest for Valerio. In fact, Valerio feels his extended representation method has a sense of poetic as in the Chinese paintings. In essence, there are some similarities and connections between those two.

溪山清远图，夏圭　In Clear View of Streams and Mountains, XIA Gui

*How have I managed to arrive where you say, when I was in another city, far far away from Cecilia, and I have not yet left it?*

*这是另外一座城市，距离切奇利雅很远，而且我还不曾走出城，怎么会来到你说的地方呢？*

*The places have mingled……*
*各地都混合起来了……*

*……Cecilia is everywhere……*
*……到处都是切奇利雅……*

*……Here, once upon a time, there must have been the meadow of the low sage……*
*……这里曾经是鼠尾草场……*

*……my goats recognize the grass on the traffic island.*
*……我的羊认出了交通安全岛那边的草。*

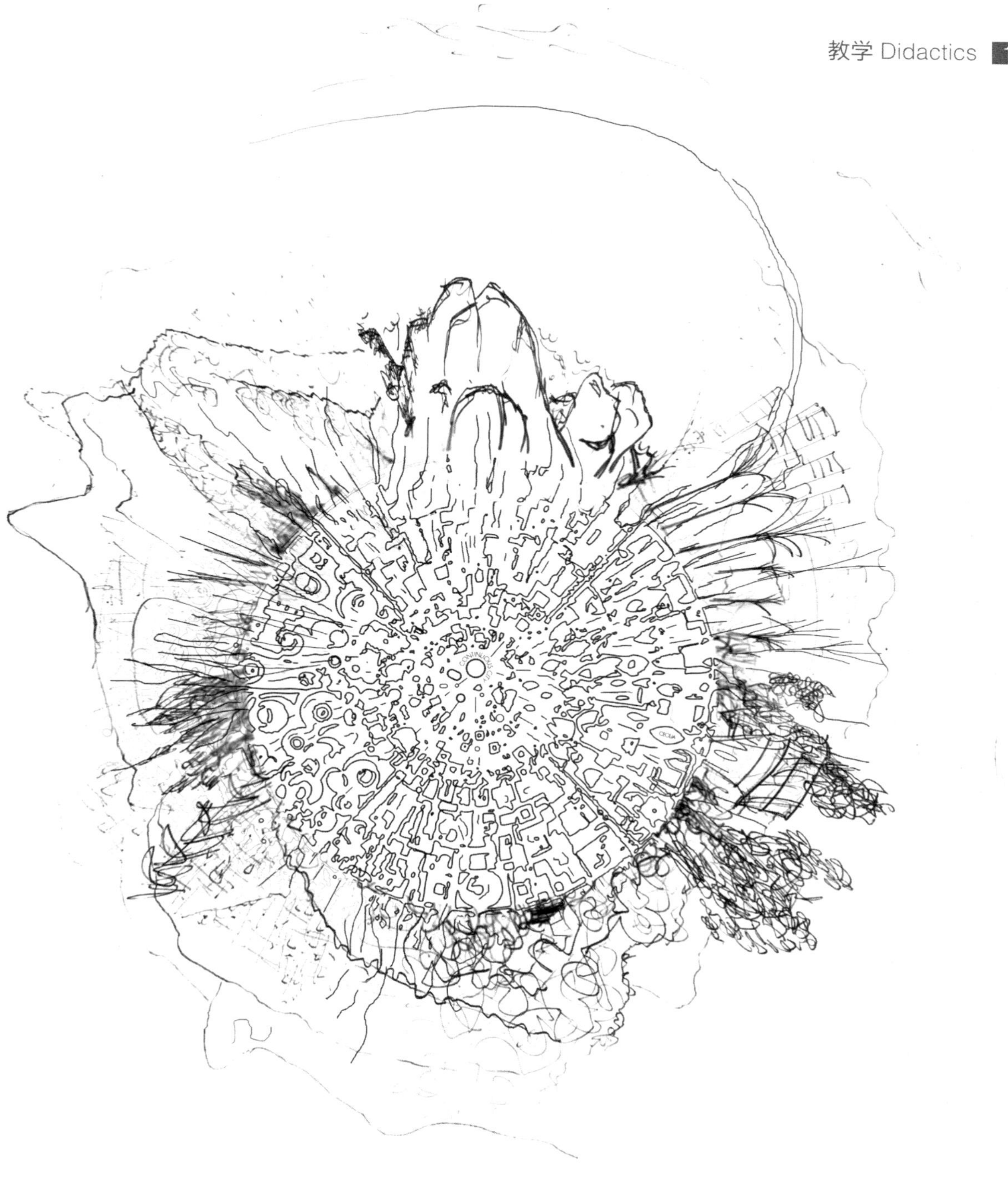

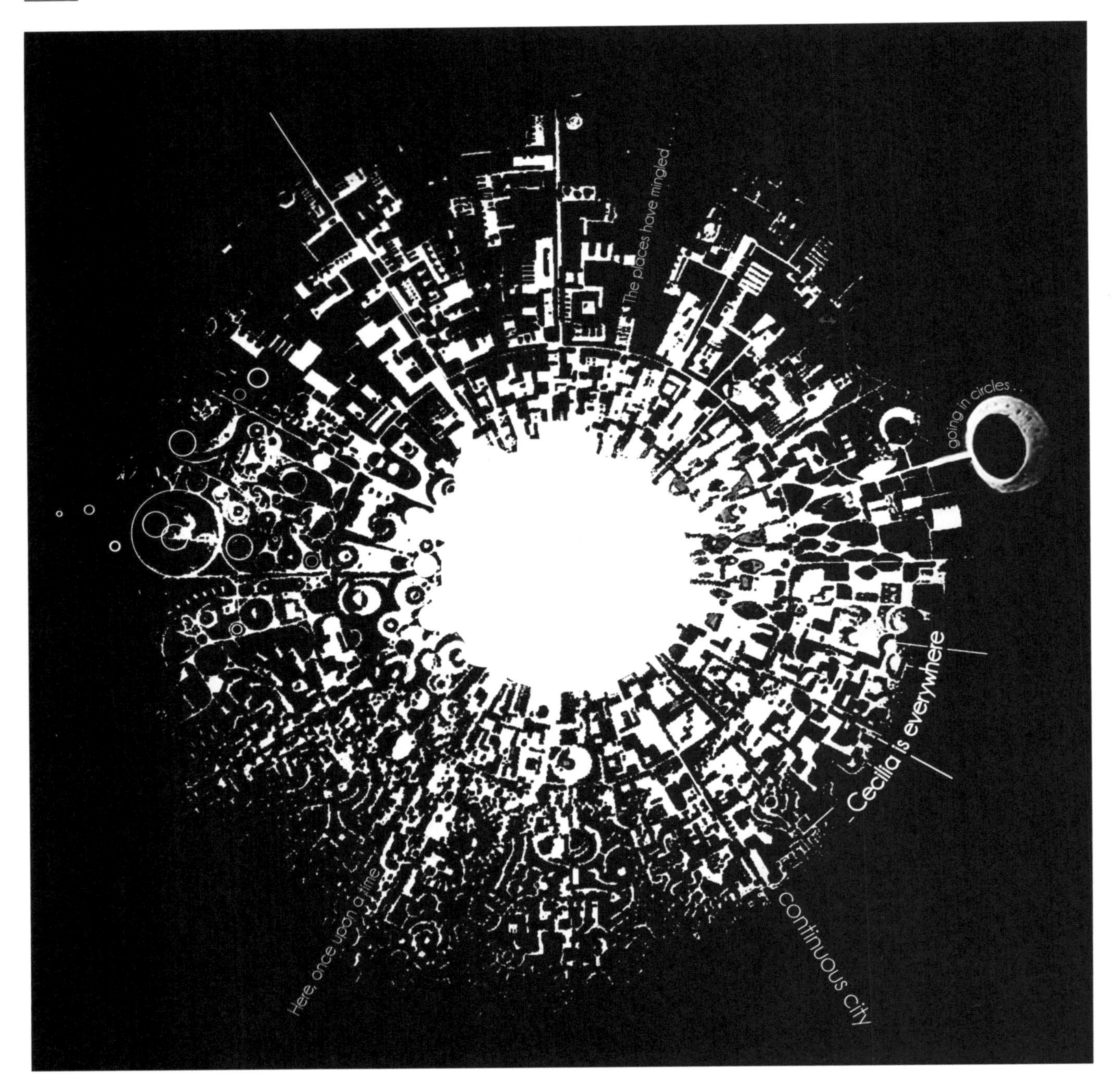
The places have mingled...
going in circles...
Cecilia is everywhere
continuous city
Here, once upon a time

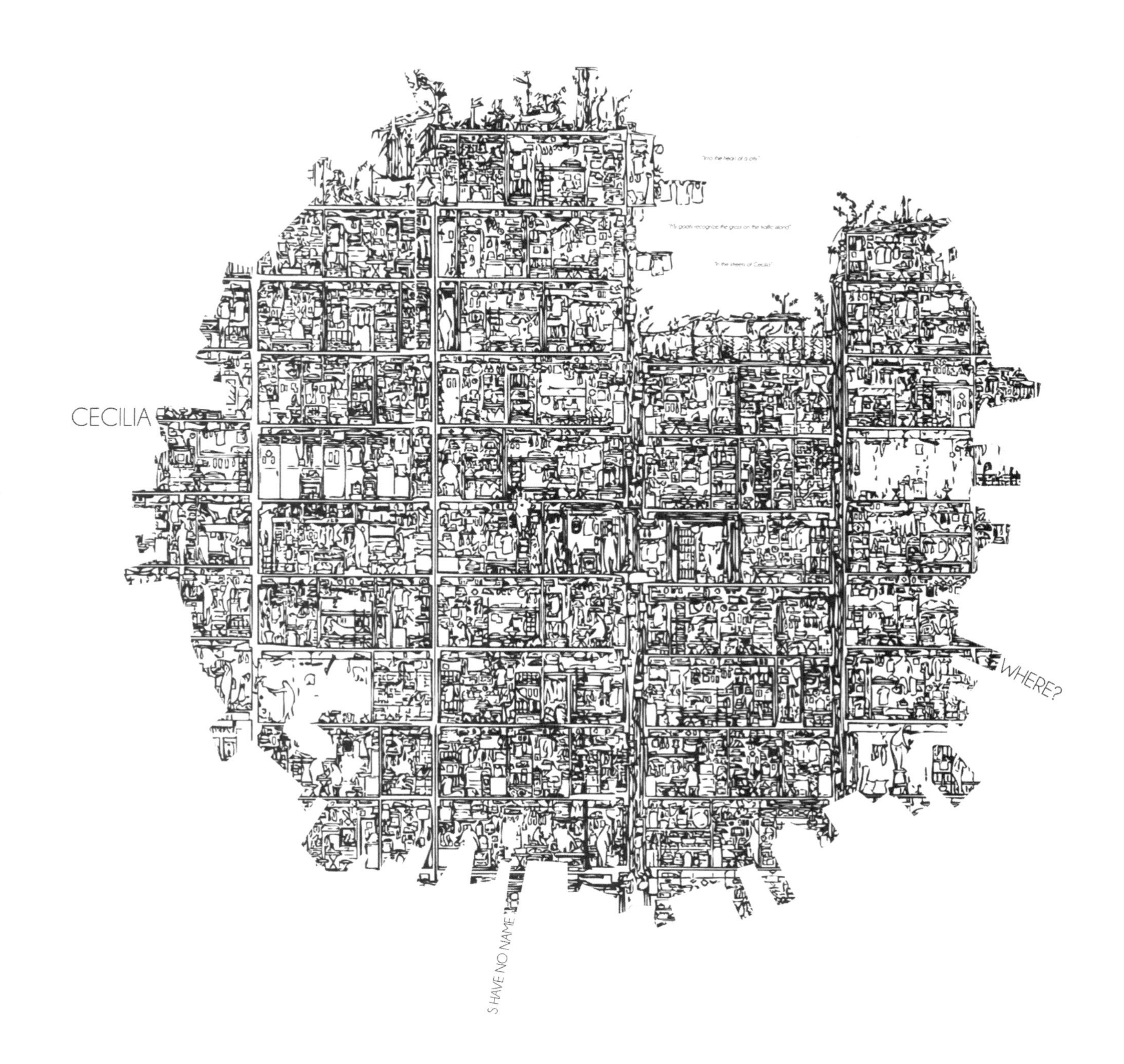
"Into the heart of a city"
"My goats recognize the grass on the traffic island"
"In the streets of Cecilia"
CECILIA
WHERE?
S HAVE NO NAME

*Laudomia's special faculty is that of being not only double, but triple; it comprehends, in short, a third Laudomia, the city of the unborn.*

*劳多米亚独特之处在于她不仅是双胞胎，而且是三胞胎，即还有第三个劳多米亚，那是尚未诞生者的城市。*

*……it is the Laudomia of the dead, the cemetery.*

*……这是死者的劳多米亚，是墓地。*

*……the pattern of the streets and the arrangement of the dwellings repeat those of the living Laudomia……*

*……街道的样式和房屋的顺序都仿照生者的劳多米亚……*

*……in both, families are more and more crowed together ……*

*……每个家庭都越来越拥挤 ……*

*……in compartments crammed one above the other.*

*……密密麻麻地重叠着。*

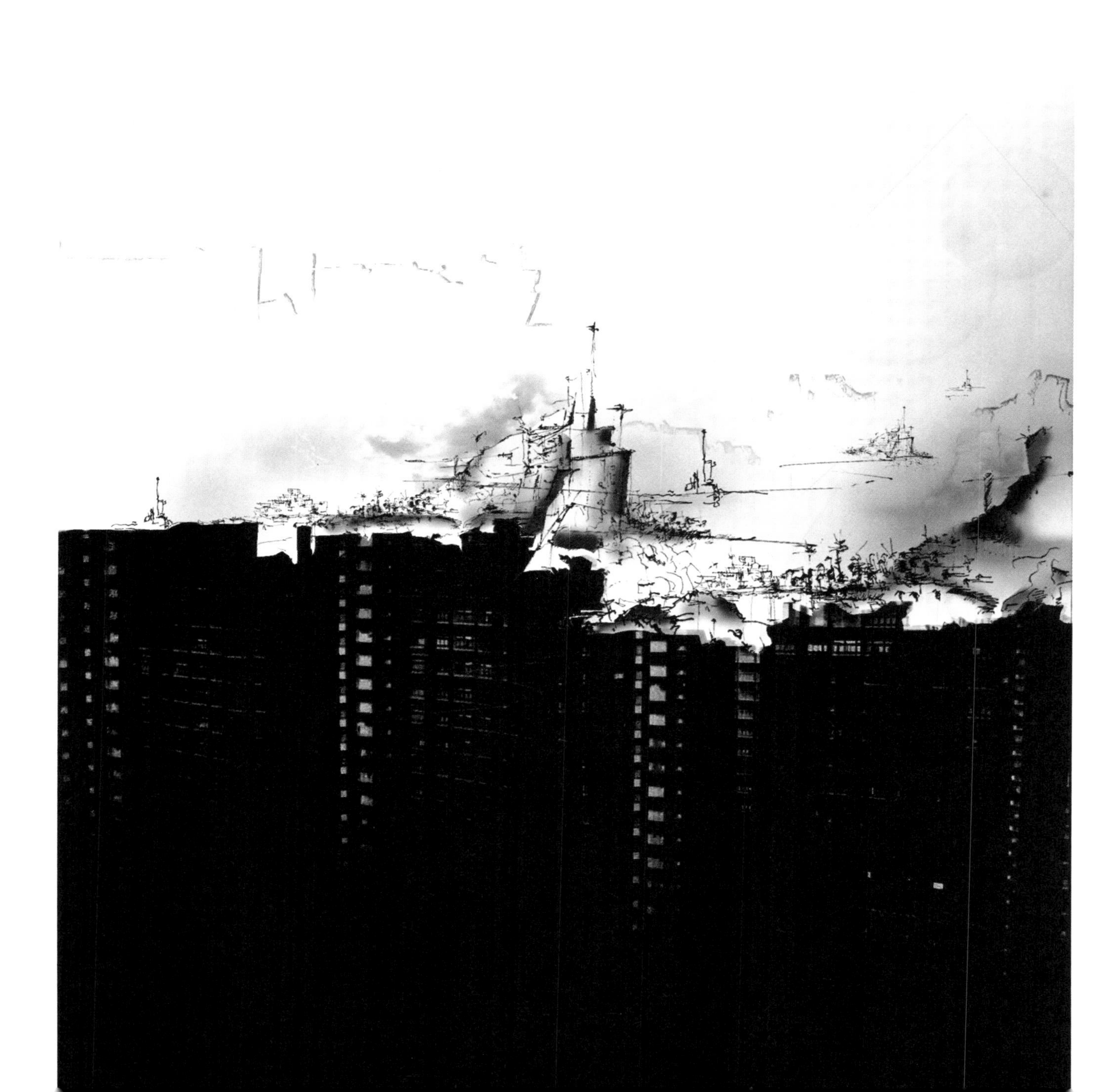

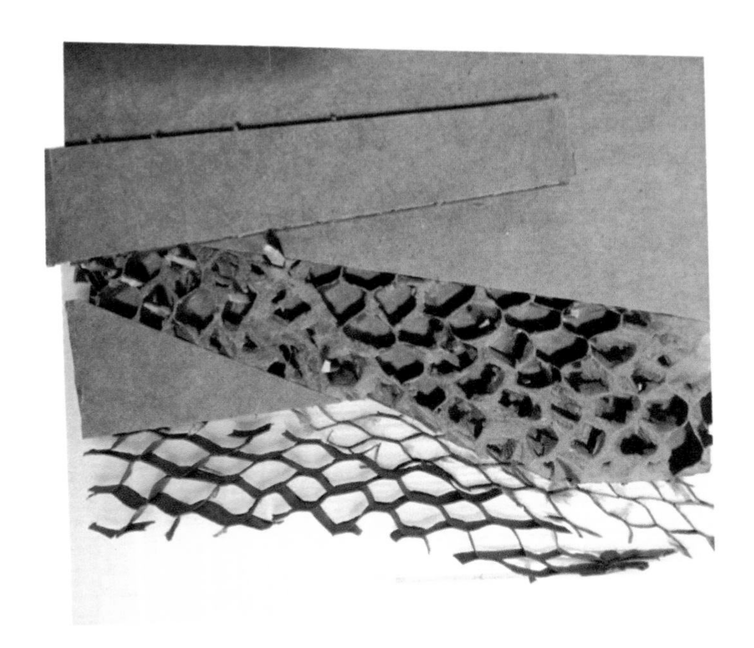

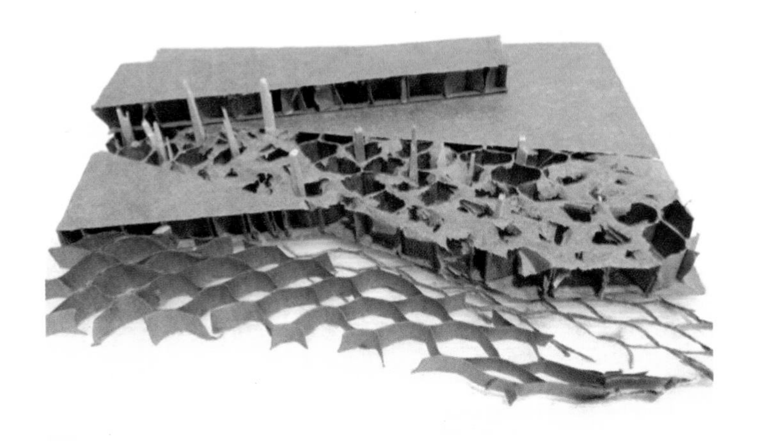

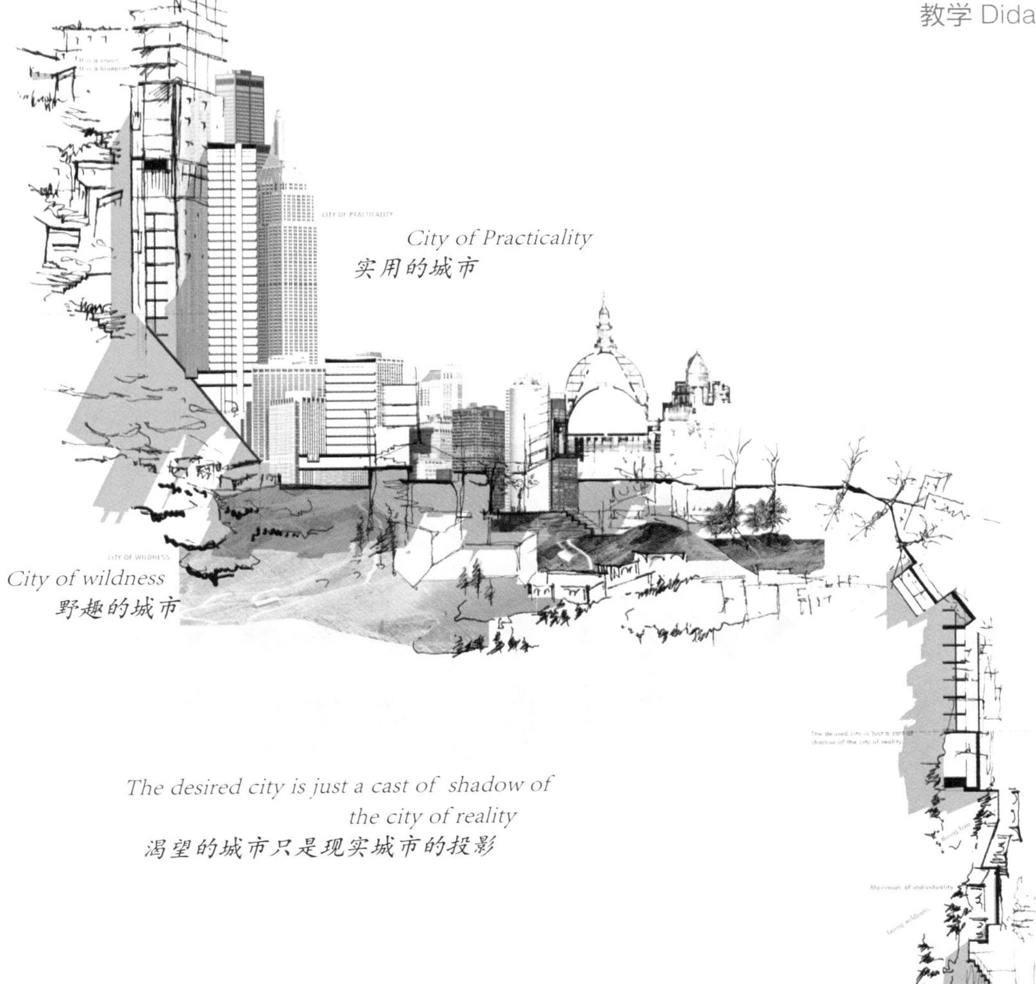
City of Practicality
实用的城市
City of wildness
野趣的城市
The desired city is just a cast of shadow of the city of reality
渴望的城市只是现实城市的投影

*Being abandoned for the practicality and functionality*

*放弃实用与功能*

*But one reaching there, it never ended up differently than the reality city*

*但你到达那里时，它结束的方式与现实城市如出一辙*

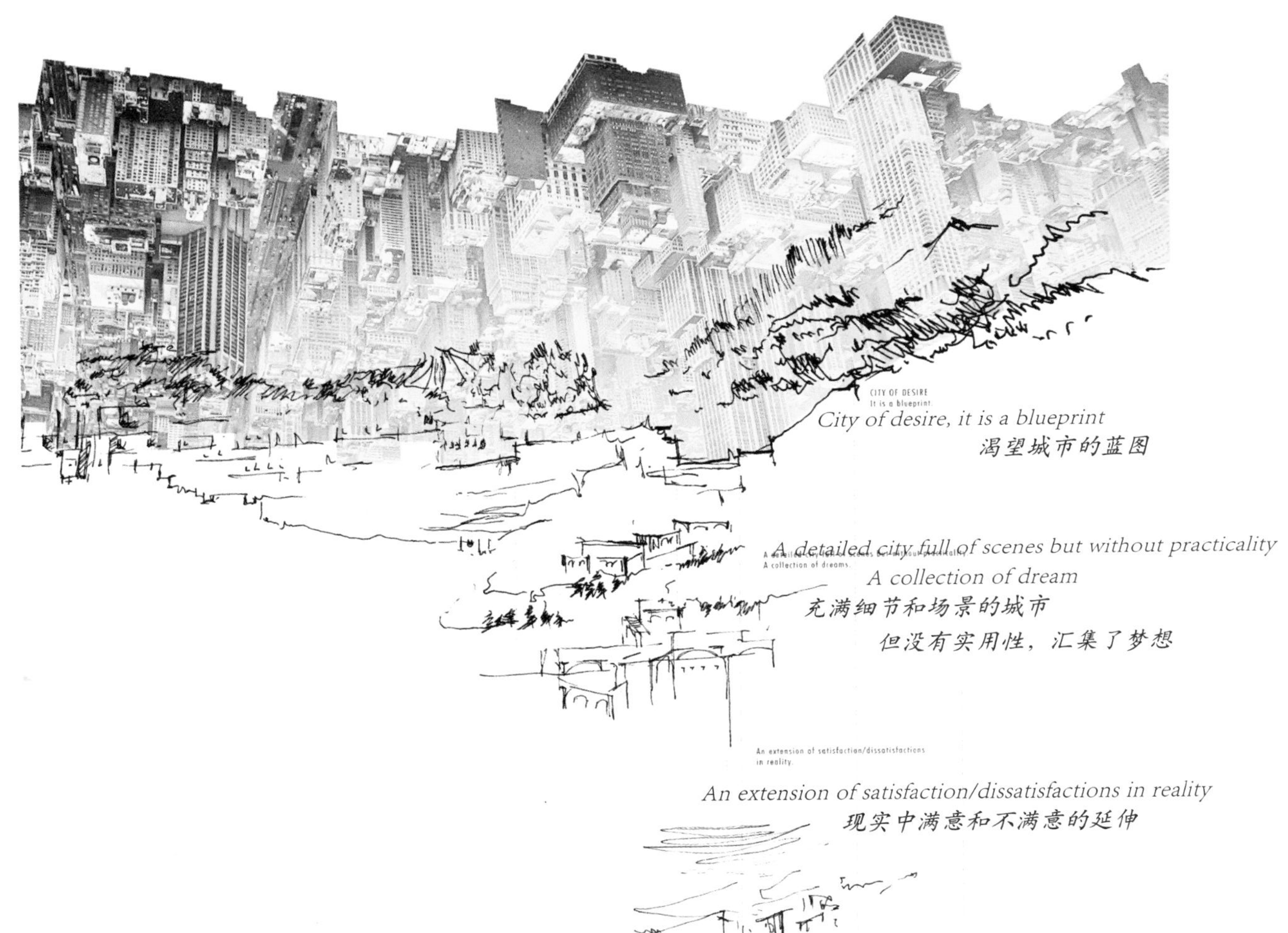


*City of desire, it is a blueprint*

渴望城市的蓝图

*A detailed city full of scenes but without practicality*

*A collection of dream*

充满细节和场景的城市

但没有实用性，汇集了梦想

*An extension of satisfaction/dissatisfactions in reality*

现实中满意和不满意的延伸

*The attempt to build up a city of desire never stops*

创造渴望城市的尝试永远不会停息

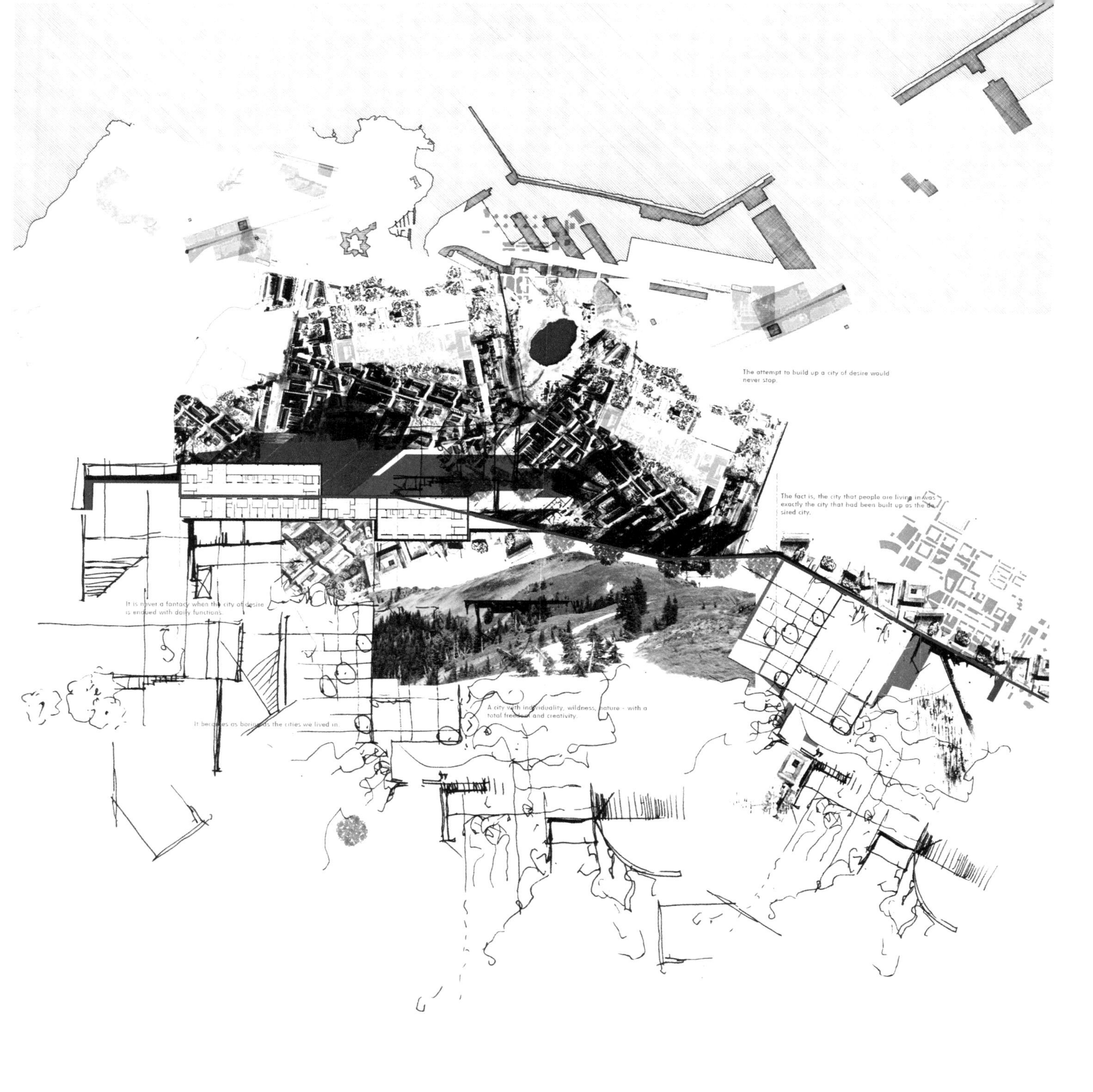
The attempt to build up a city of desire would never stop.
The fact is, the city that people are living in was exactly the city that had been built up as the desired city.
It is never a fantacy when the city of desire is endued with daily functions.
A city with individuality, wildness, nature - with a total freedom and creativity.
It becomes as boring as the cities we lived in.

*From one part to the other, the city seems to continue, in prospective, multiplying its repertory of images……*

*从这面到那面，城市的各种形象在不断翻番……*

*……you have only to walk in a semicircle and you will come into view of* **Moriana***'s hidden face……*

*……只要绕半个圈子，你就会看到***莫里亚纳***掩饰着的另一幅面孔……*

*……but instead it has no thickness, it consists only of a face and an obverse……*

*……但是却没有厚度，只有正反两面……*

*……like a sheet of paper……*
*……with a figure on either side……*

*……就像一张两面都有画的纸……*

*……which can neither be separated nor look at each other.*

*……两幅画既不能分开，也不能对看。*

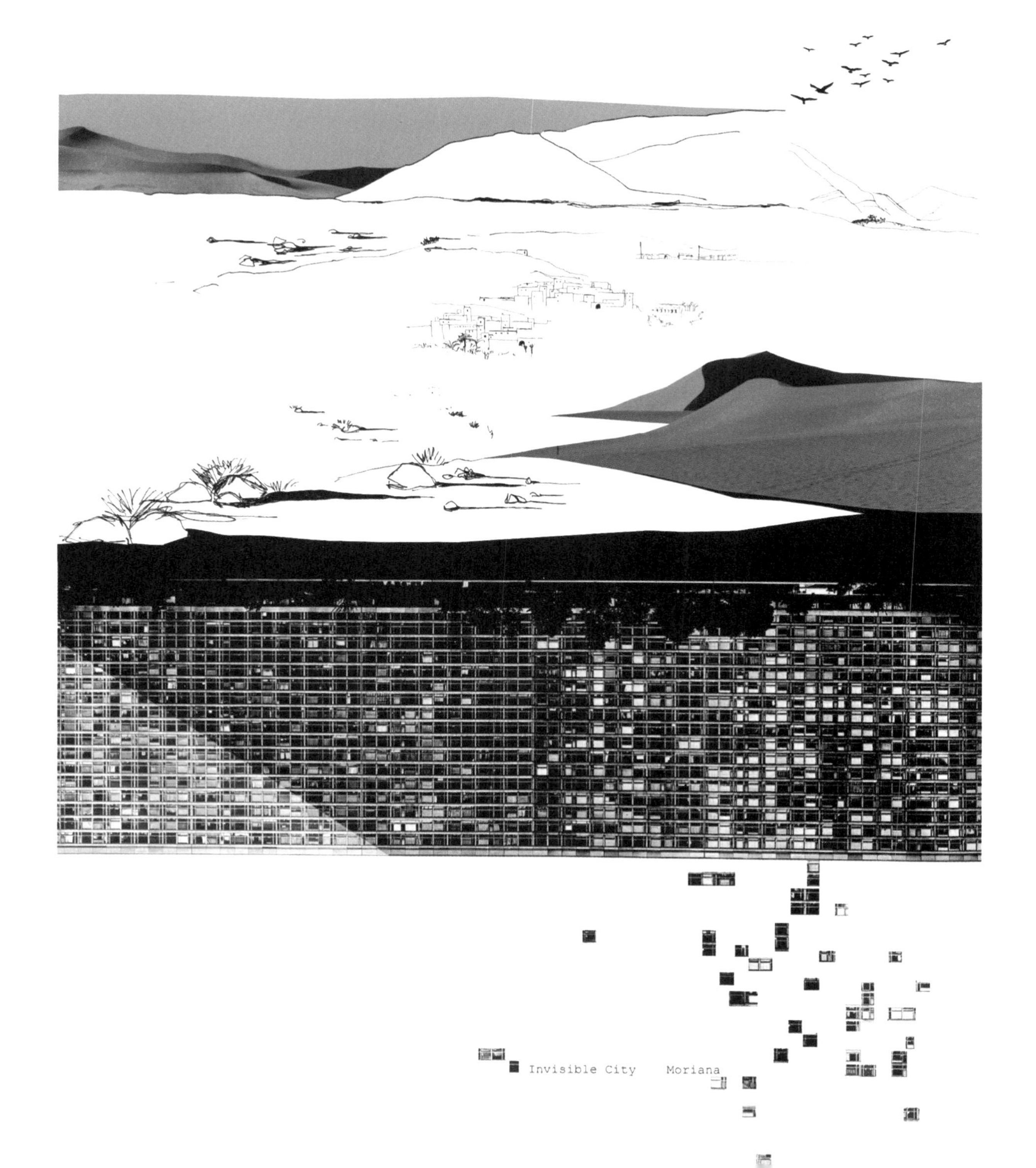
Invisible City
Moriana

Invisible City Moriana

*As soon as I set foot there, everything I had imagined was forgotten……*
*一踏上这块土地，我就立即忘掉了以前的所有想象……*

*……Pyrrha had become what is Pyrrha;*
*……皮拉变成了皮拉自己的样子。*

*From that moment on the name Pyrrha has brought to my mind this view, this light, this buzzing, this air in which a yellowish dust flies: obviously the name means this and could mean nothing but this.*
*从那以后，皮拉这个名字在我脑海唤起的就是这副景象，这种光线、这种嗡嗡的声音、这种黄尘浮动的空气。很显然，除此之外，这个名字不可能具有其他意义。*

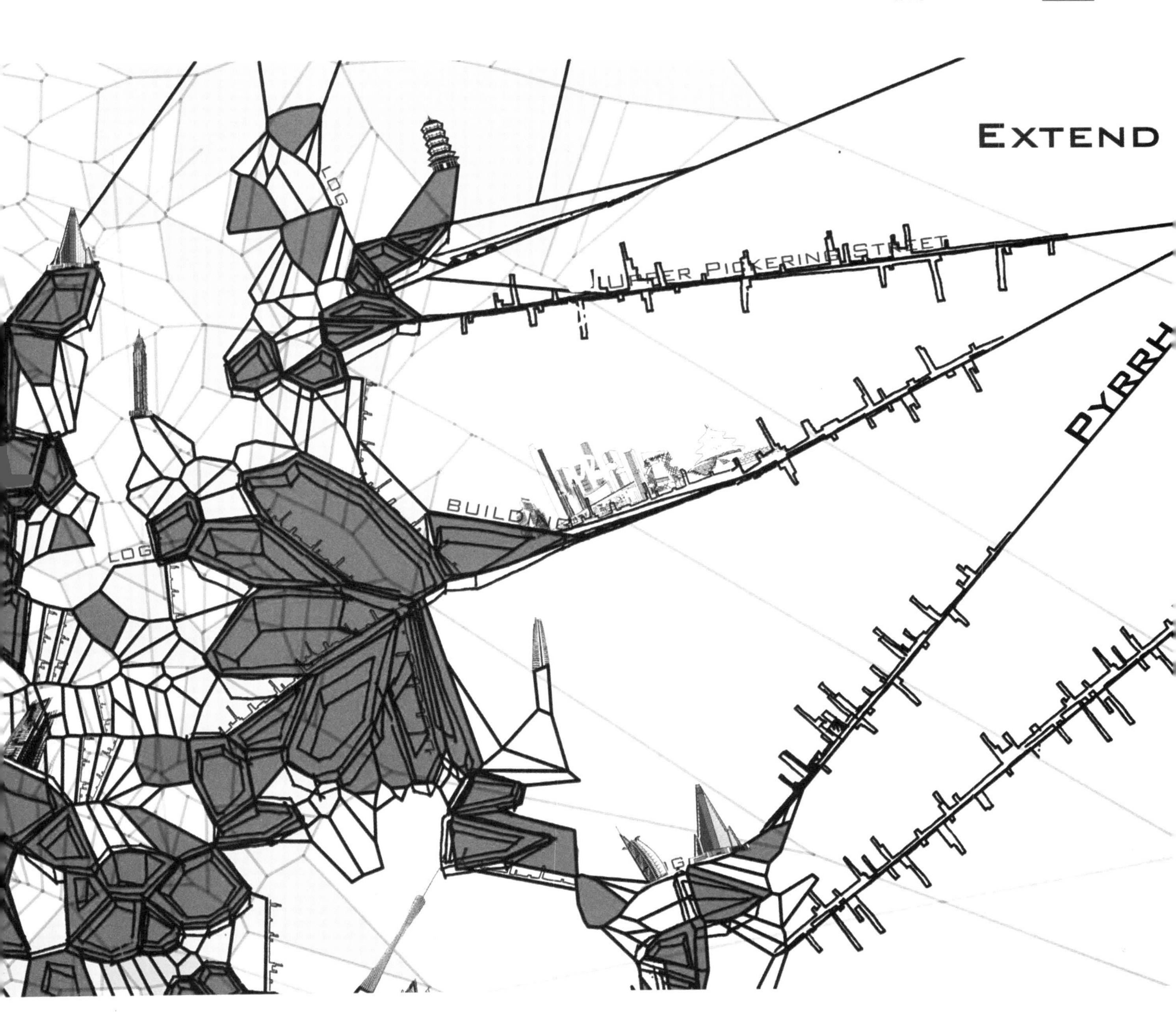
EXTEND
PICKERING
PYRRH
LOG
LOG

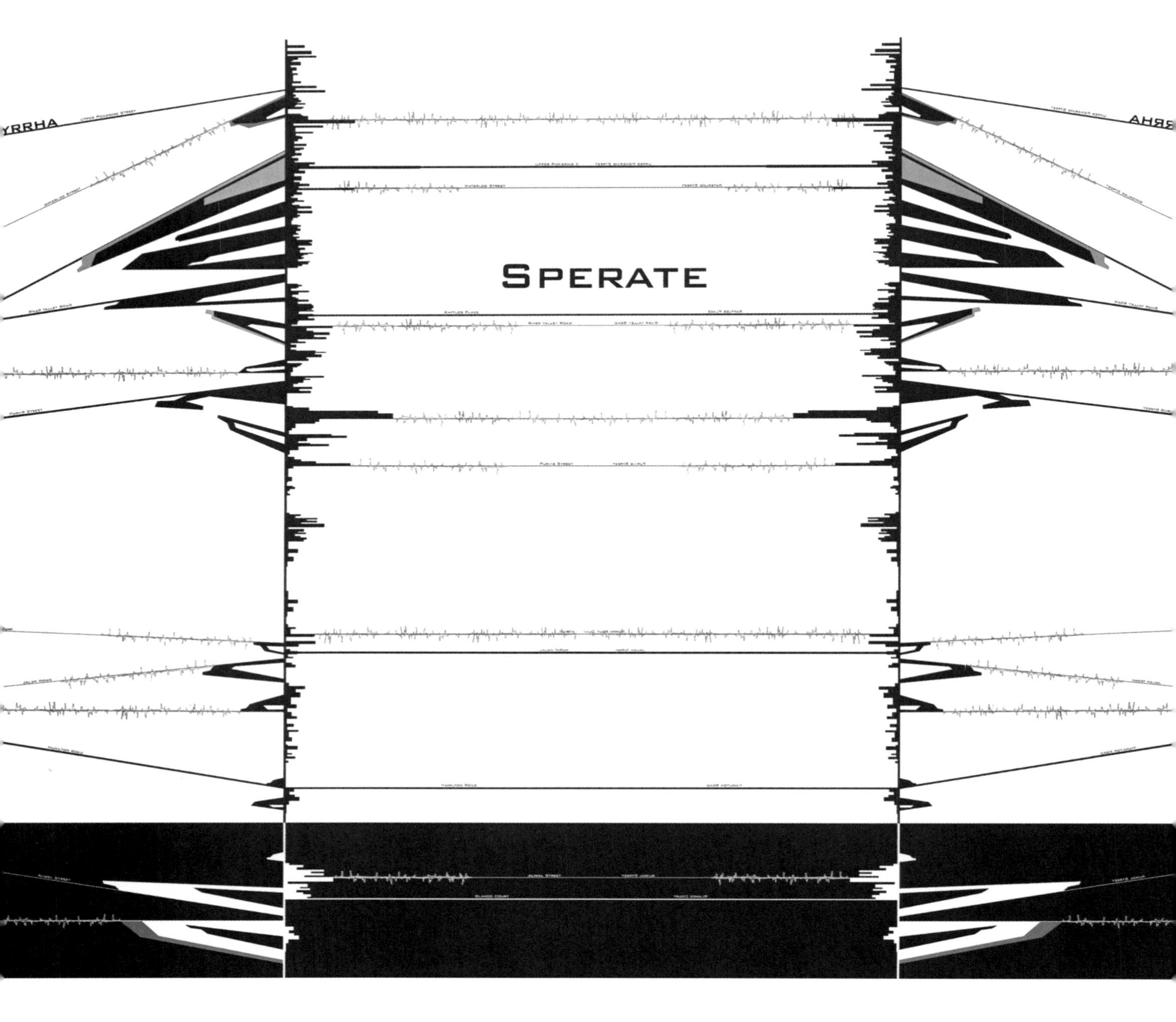
SPERATE

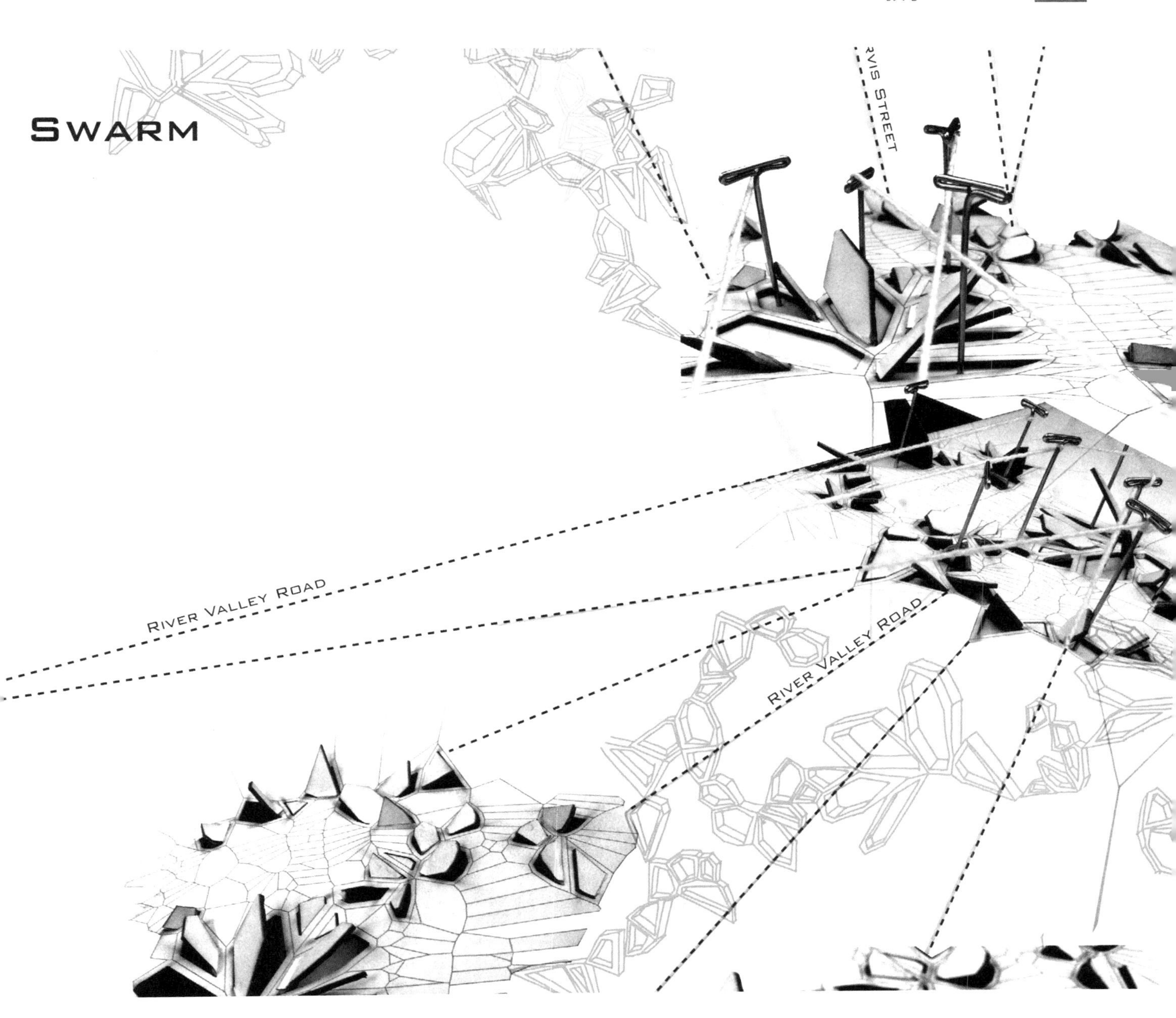
SWARM
RVIS STREET
RIVER VALLEY ROAD
RIVER VALLEY ROAD

of city

*Irene is a name for a city in the distance, and if you approach, it changes.*

*伊莱那是一座从远方看到的城市的名字，如果走进她，她就变了。*

CITY 7 you approach

fire crackers: light of festivals

CITY 6 you approach

CITY 5 standing in its midst

concentrated window

*There is the city where you arrive for the first time;*

*and there is another city which you leave never to return.*

*人们初次抵达的时候，城市是一种模样；而永远离别的时候，她又是另一种模样。*

CITY 4 leave never to return

CITY 3 arrive for the first time

plateau

CITY 2 trapped by it and never leave

roads winding down to the valley

CITY 1 who pass it without entering

*Each deserves a different name; perhaps I have already spoken of Irene under other names;*

*perhaps I have spoken only of Irene.*

*每个城市都该有自己的名字；也许我已经用其他名字讲过伊莱那；也许我讲过的那些城市都只是伊莱那。*

shepherd on plateau
roads winding down to the valley
first light through the plateau
music of bass drums and trumpets
concentrated window
Shadows of the garden
Alleys of in the city
towers of signiture fires
towers of signiture fires
skycrapers
Alleys of in the city
PROJECTION OF INVISIBLE CITY
IRENE
rene is the city visible when you lean
out from the edge of the plateau at
the hour when the lights come on,
and in the limpid air, the pink of the
settlement can be discerned spread
out in the distance below: where the
windows are more concentrated,
where it thins out in dimly lighted al-
eys, where it collects the shadows
of gardens, where it raises towers
with signal fires; and if the evening is
nisty, a hazy glow swells like a milky
ponge at the foot of the gulleys.

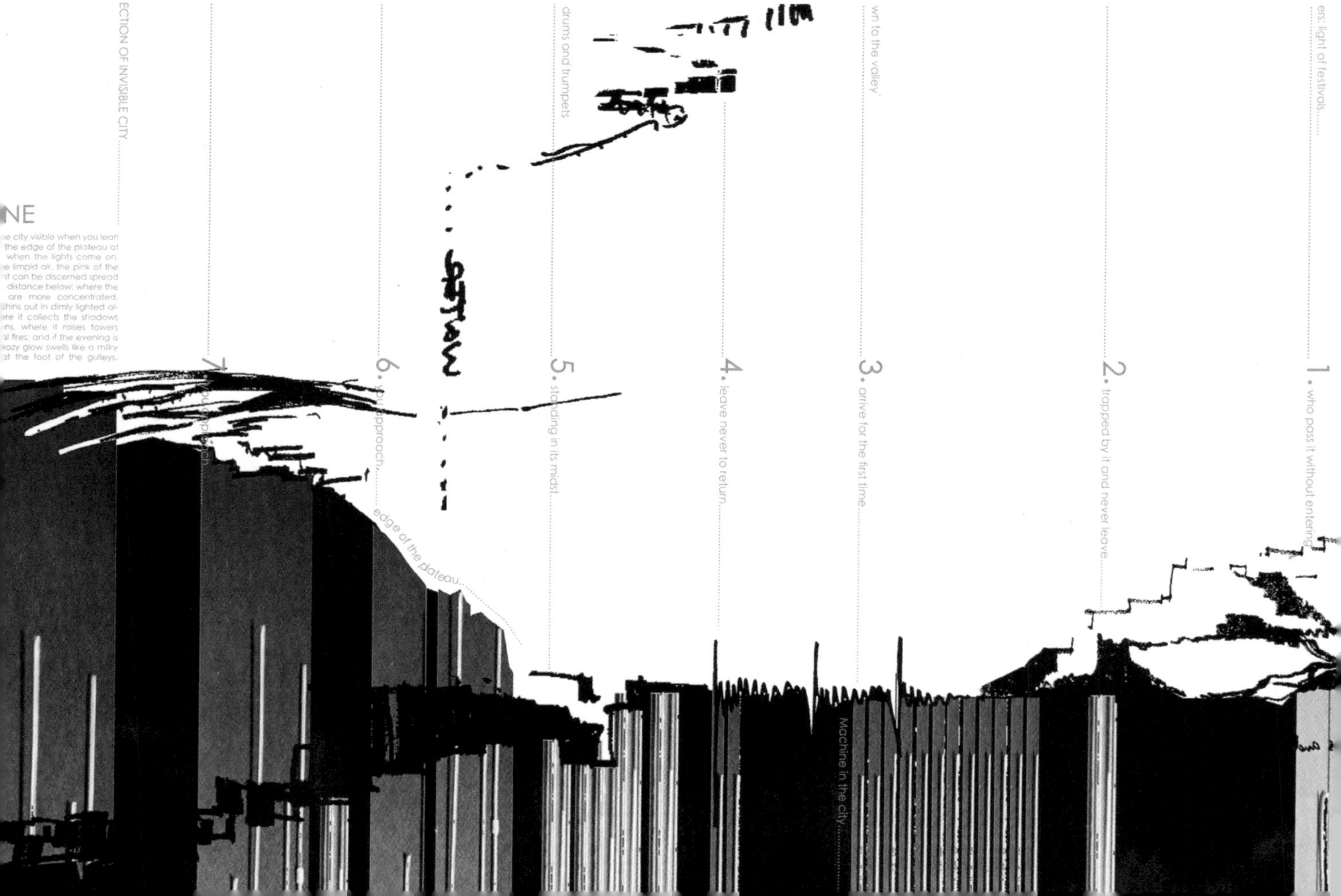

ECTION OF INVISIBLE CITY
NE
e city visible when you lean the edge of the plateau at when the lights come on, e limpid air, the pink of the nt can be discerned spread distance below: where the are more concentrated, thins out in dimly lighted al- ere it collects the shadows ns, where it raises towers al fires; and if the evening is hazy glow swells like a milky at the foot of the gulleys.
7.
6. you approach
5. standing in its midst
drums and trumpets
4. leave never to return
3. arrive for the first time
wn to the valley
2. trapped by it and never leave
1. who pass it without entering
ers: light of festivals
edge of the plateau
Machine in the city

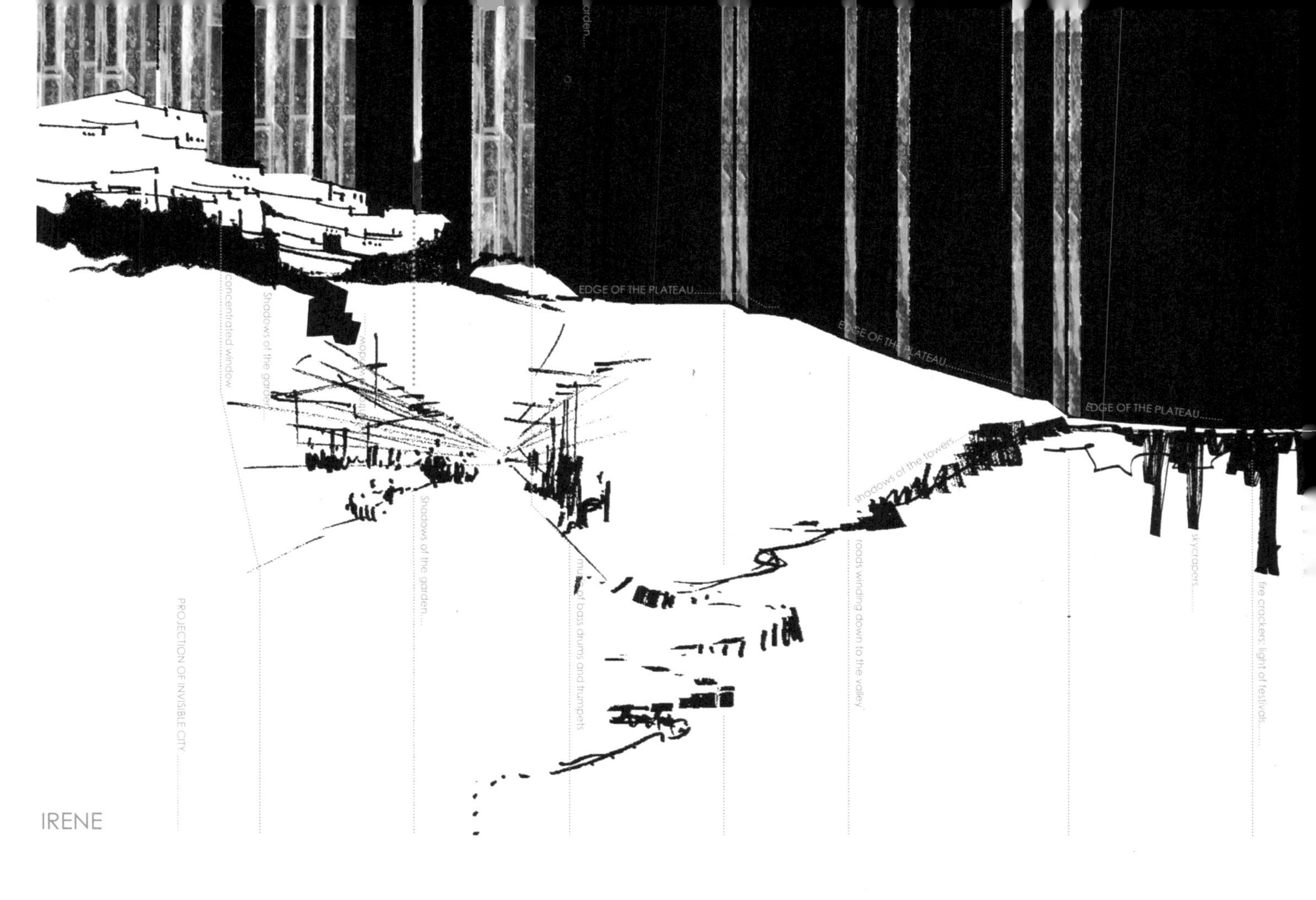

EDGE OF THE PLATEAU........
EDGE OF THE PLATEAU........
EDGE OF THE PLATEAU........
concentrated window
Shadows of the garden......
Shadows of the garden......
shadows of the towers......
roads winding down to the valley
skycrapers......
fire crackers; light of festivals......
of bass drums and trumpets
PROJECTION OF INVISIBLE CITY......
IRENE

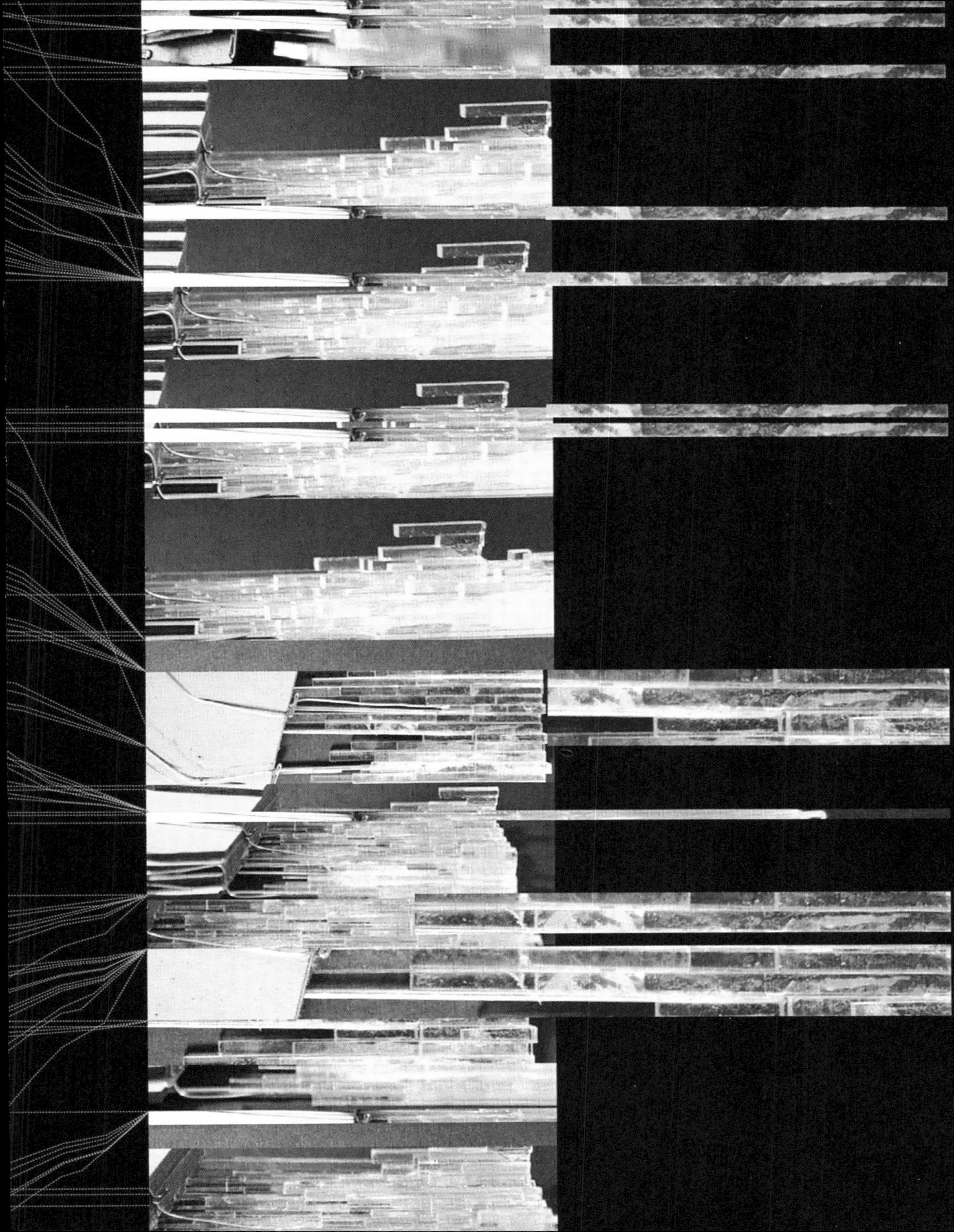

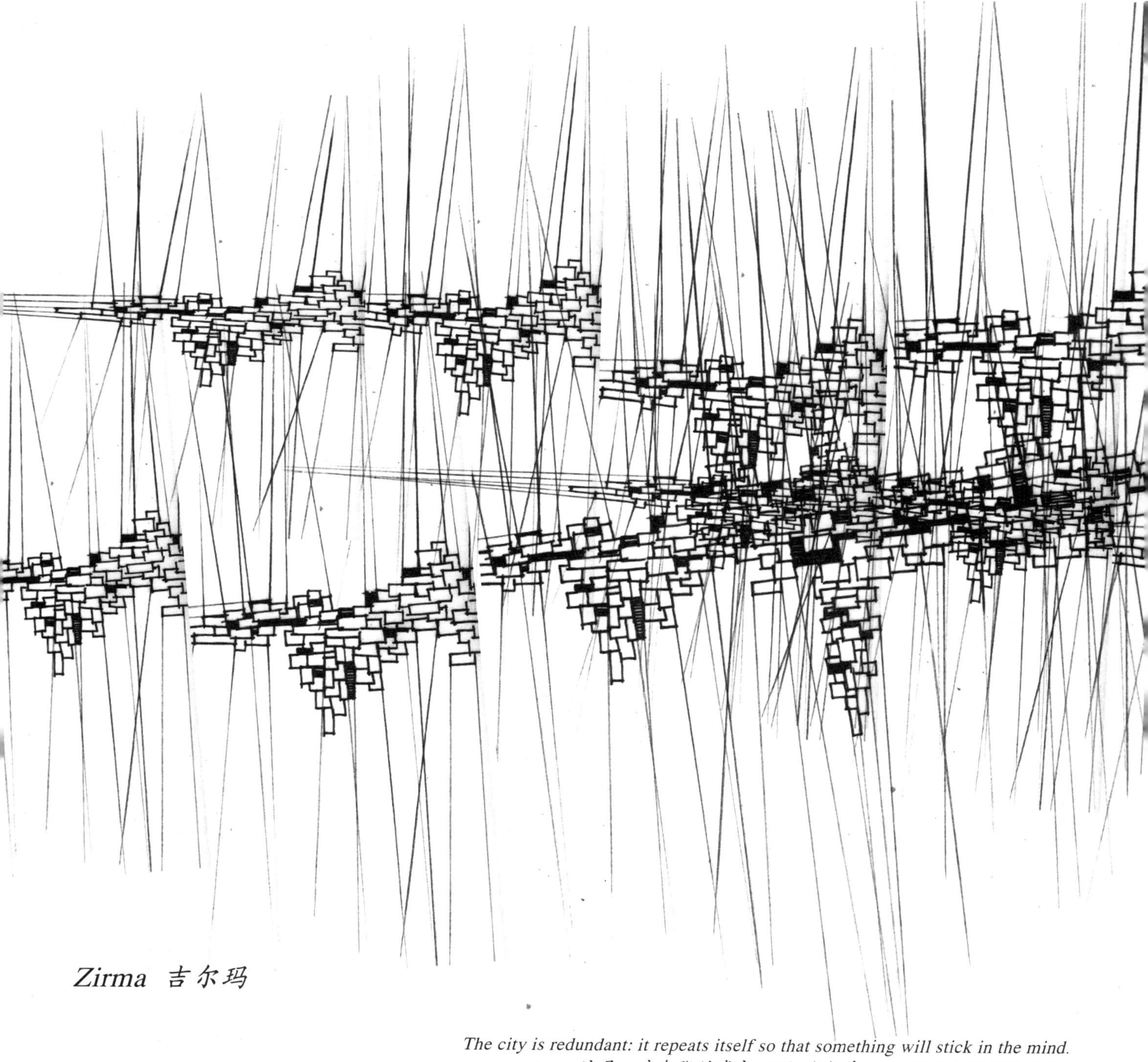

*Zirma* 吉尔玛

*The city is redundant: it repeats itself so that something will stick in the mind.*
这是一座夸张的城市：不断重复着一切，好让人们记住自己。

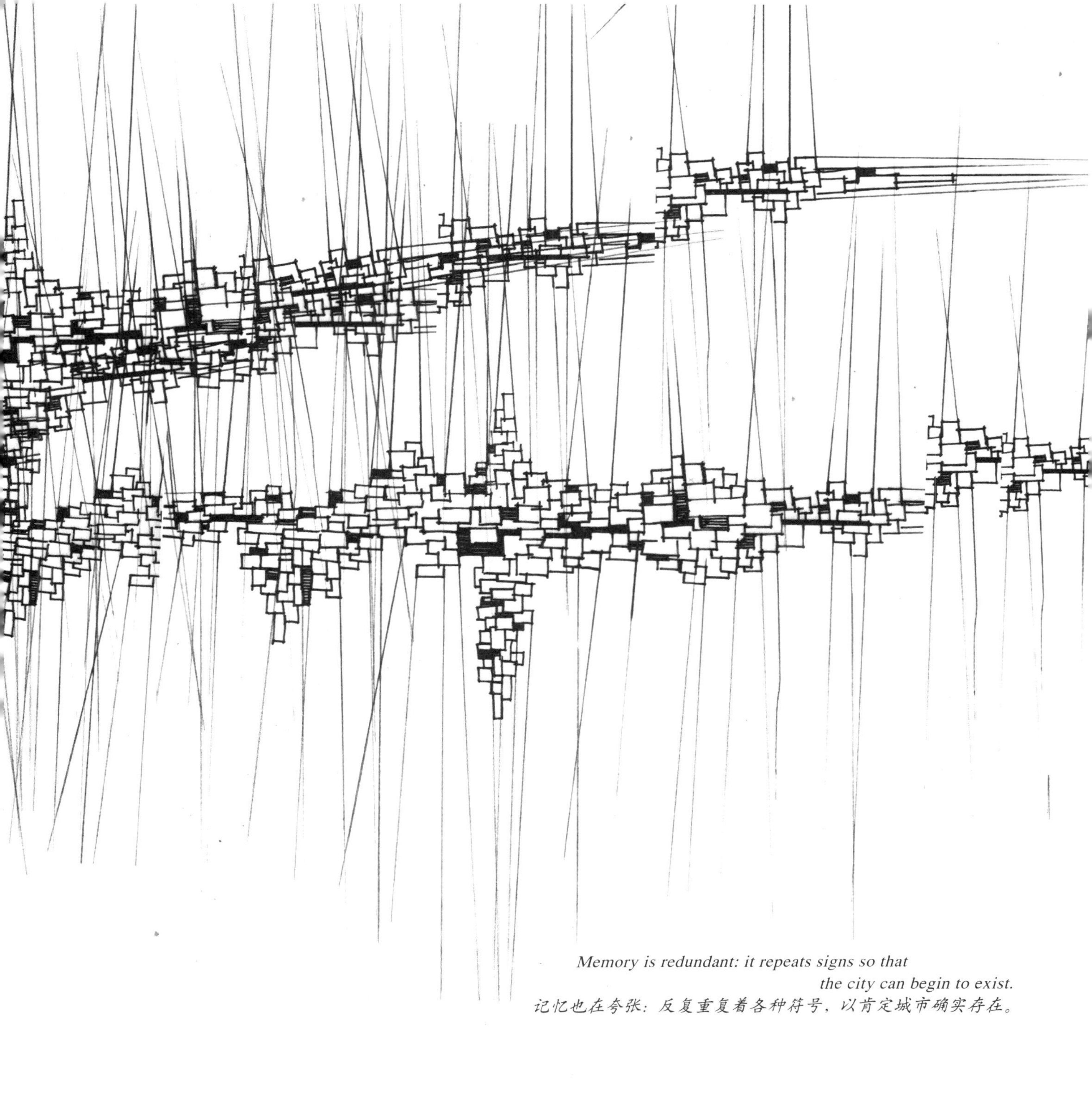

*Memory is redundant: it repeats signs so that the city can begin to exist.*

记忆也在夸张：反复重复着各种符号，以肯定城市确实存在。

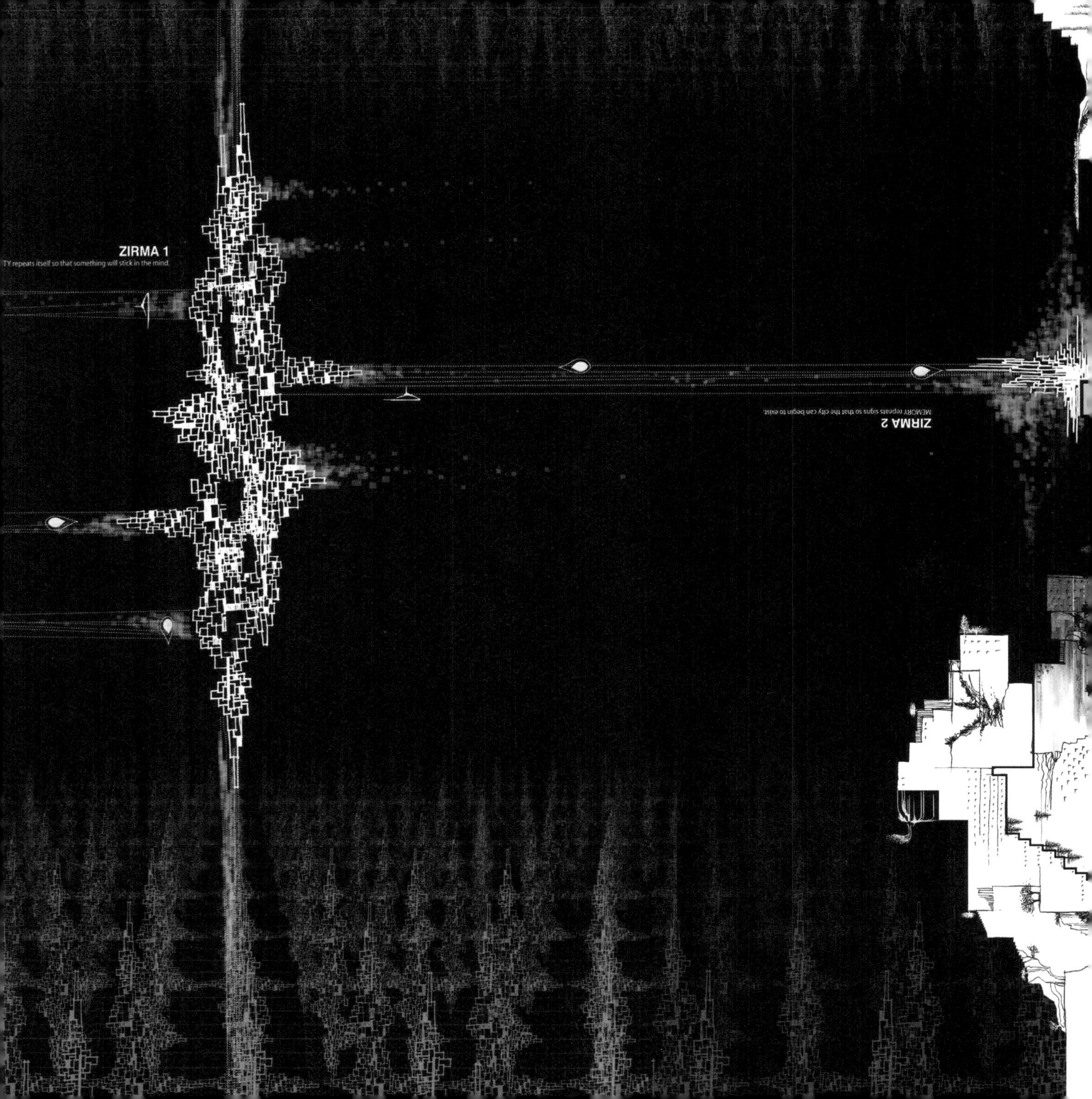
ZIRMA 1
TY repeats itself so that something will stick in the mind.
ZIRMA 2
MEMORY repeats signs so that the city can begin to exist.

*Happy the man who has* **Phyllis** *before his eyes each day and who never ceases seeing the things it contains.*

能够每天都看到**菲利斯**所包含的看不完的景致的人，他们是多么幸福啊！

*Phyllis is a space in which routes are drawn between points suspended in the void.*

菲利斯是一个空间，虚无中各点之间都连着通道。

*Many are the cities like Phyllis, which elude the gaze of all, except the man who catches them by surprise.*

*像菲利斯这样的城市很多，它们能够躲过所有凝视的目光，却躲不过那些出其不意投来的目光。*

# 美国景观设计师协会荣誉奖
# ASLA Honor Award

*不同领会，不同思考，不同绘画*

——陈巍，毕业于宾夕法尼亚大学景观系

*See Different, Think Different, Draw Different*

—— Wei Chen, Student ASLA, Graduate, University of Pennsylvania

*这些是我从“表现话题：景观绘图”的课程中学习并获得的关于景观表现的新想法和技术。这些新的景观表现方法的目的在于结合模型、绘画、照片、电影以及电脑图形技术以更好地提炼场地、文化、物理特征、形态的本质。*

*These are new ideas and techniques of landscape representation I acquired through study of the course "Topics in Representation: Landscape Drawing". These new landscape representation methods aim at combining advantages of modeling, drawings, photos, movies, as well as computer graphic techniques in order to better capture the essence of a site, its culture, its physical characteristic, its morphology.*

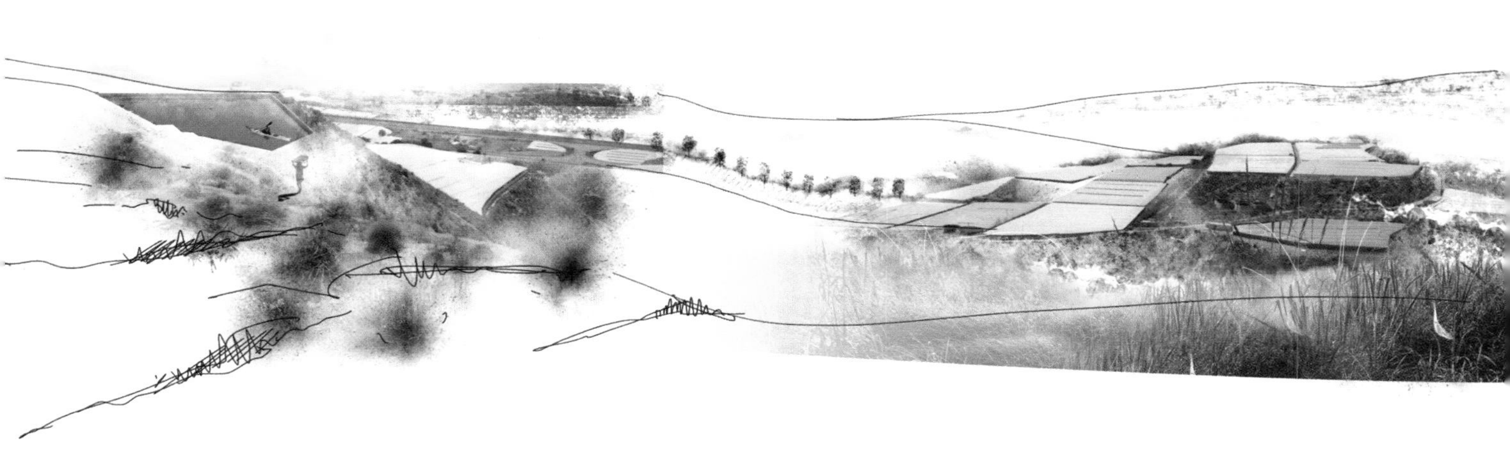

手法:相片拼接、Photoshop笔刷处理、绘图板手绘

思路:阿尔武尼奥尔当地农业资源丰富,自然风光优美。该作品旨在突显场地最具价值的景观元素,并将这些元素串联在一起,形成一首完整的“叙事诗”:自山顶俯瞰延绵山脉、海岸线,顺蜿蜒的山路而下,在山坳里荡舟,在山谷间瓜果芳香中徜徉,不知不觉中来到广袤的地中海边……将对场地的美好想象转化为纸笔间一个个特写镜头,其间留白更丰富了观者的想象。

Method: *Photo montage, Photoshop brush handling, freehand sketch with drawing board*

Idea: *Albunol is a place with abundant agricultural resources and a beautiful natural scenery. This drawing aims to highlight the most valuable landscape elements of the site, and it aims to connect them together to compose an "epic": from the top of the mountain overlooking the continuous mountain ranges and coastlines, along the winding path, boating in the col, strolling in the valley full of fruit fragrances, unconsciously arriving at the vast Mediterranean coast... Gorgeous imagination is painted as a series of close-up shots, the white space in between leaves the space for the viewers to imagine.*

手法：模型制作、相片拼接、Photoshop笔刷处理、绘图板手绘

思路：摩洛哥北部小城沙温是个充满梦幻色彩的城市。该作品将城市不同场景、物件拼接，以展现沙温错综的街巷以及独特的文化。创作过程首先是通过空间想象制作实体模型，接着基于实体模型的空间布局进行相片拼接。

Method: *Modeling, Photo montage, Photoshop brush handling, freehand sketch with drawing board*

Idea: *Chefchaouen, located in northern Morocco, is a town full of fantasy colors. This drawing is a montage of different scenes and objects in order to represent the intricate lanes and the unique culture of Chefchaouen. The process of this work is, at first, the making of the physical model according to spatial imagination, and then the joining of the photos together, based on the spatial layout of the physical model.*

手法：电影回想、模型制作、相片剪辑、Photoshop 笔刷处理、绘图板手绘

思路：科幻电影《银翼杀手》中镜头在城市间游走的画面给我留下了极深的印象。通过对电影画面的回想及想象，我首先制作了与电影中场景类似的空间实体模型，接着搜集影像素材，并根据模型空间组合形式对素材进行拼接。

Method: *Memory from movie, modeling, Photo montage, Photoshop brush handling, freehand sketch with drawing board*

Idea: *The scenes about the movement in the city from the science fiction film "Blade Runner" left me a deep impression. From the memory and imagination about the scenes, at first, I made a physical spatial model similar to the scenarios in the movie, and then, I collected images, materials, and joined them together according to the space combinations of the model.*

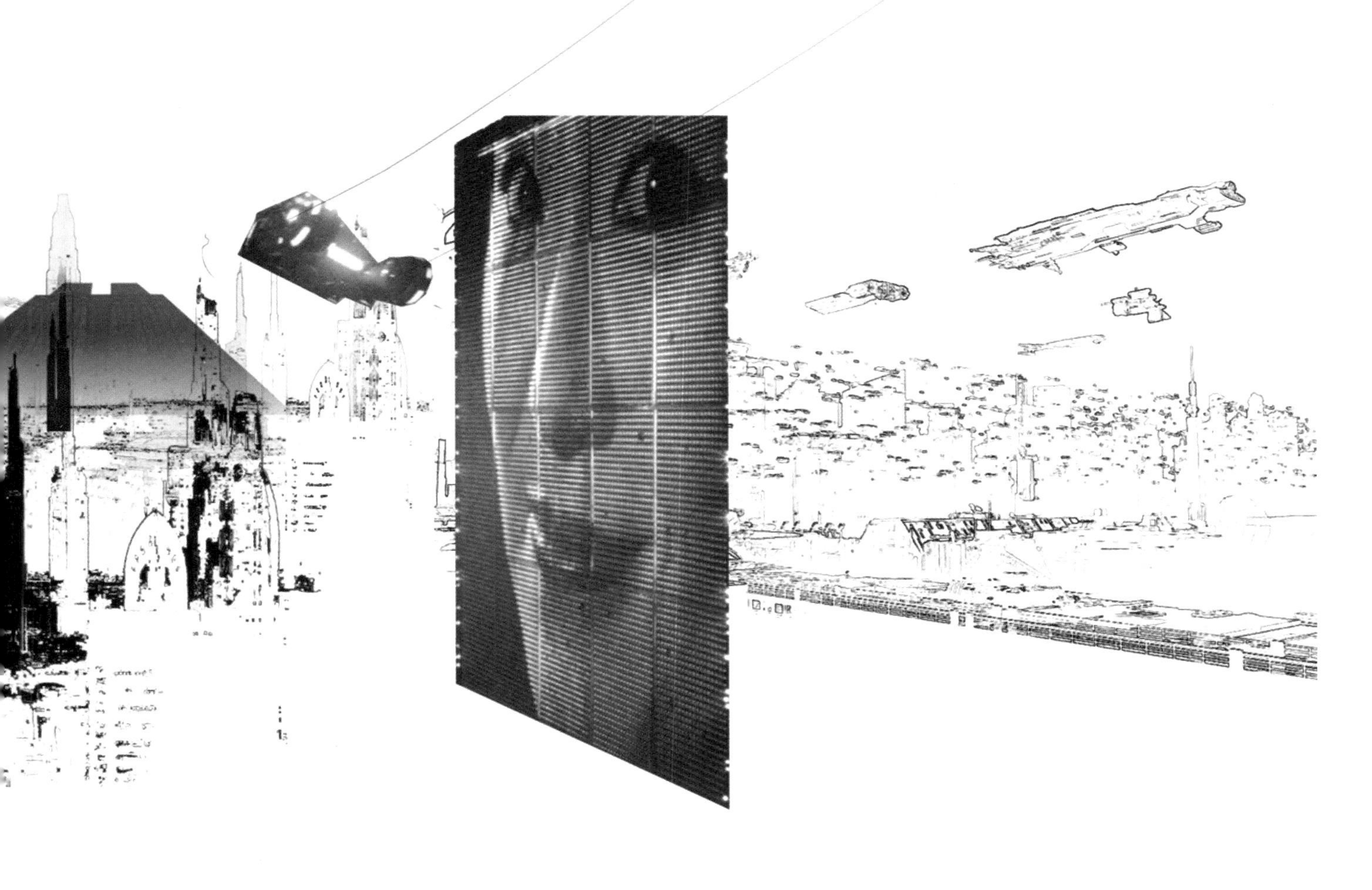

手法：文章阅读、模型制作、模型相片 Photoshop 处理、素材与模型相片拼接

思路：相比电影和音乐，文字往往更能激发人们的想象。该作品同样是基于模型制作和相片拼接。而灵感来自于作家伊塔洛·卡尔维诺在《看不见的城市》中关于左拉的描述。左拉或许只是一个存在于作家笔下的城市，却被描述得栩栩如生："左拉坐落在六河三山之间，是一座令人过目不忘的城市……左拉的秘密在于，当你的目光滑过它，就好像滑过一首完整乐曲中的不同音符……铜钟挨着理发店的条纹遮雨棚，旁边是九眼的喷泉，天文台的玻璃塔楼，卖瓜的摊贩……"

Method: *Reading, modeling, modifying model pictures with Photoshop, photo montage of materials and model*

Idea: *Compared to films and music, words tend to be more powerful to ignite stimulating imagination. Similar to the previous work, this is also based on the montage of the model and the photos. However, the inspiration derived from a story of a city named Zora, from the book "Invisible Cities", written by the writer Italo Calvino. Zora is a city that exists through the writer's pen, through which it is described vividly: "Beyond six rivers and three mountain ranges rises Zora, a city that no one, having seen it, can forget……Zora's secret lies in the way your gaze runs over patterns following one another as in a musical score……the copper clock follows the barber's striped awning, then the fountain with the nine jets, the astronomer's glass tower, the melon vendor's kiosk……"*

PAN DI ZUCCHERO
PAN DI ZUCCHERO
PORTO FLAVIA
PORTO FLAVIA

# 实践 Practice

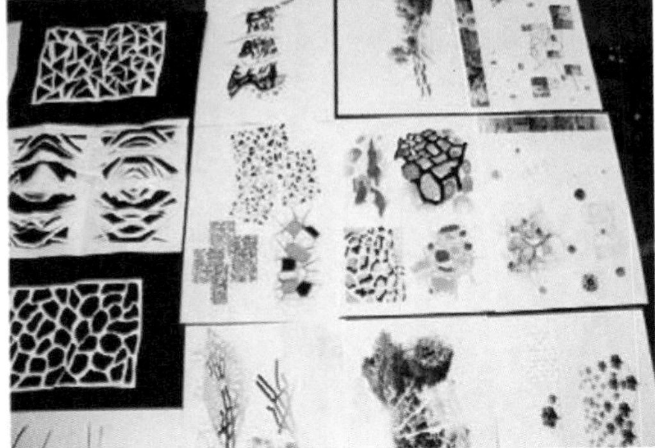

# 瓦莱里奥工作室作品

# Works from Valerio's Studio

景观的延伸表现也可以被应用到实际的设计项目中的。确实，它远不止是一种表现，本质上说，它是一种设计过程。瓦莱里奥的表现和设计相互影响着：他事务所 APScape 的专业实践孕育着延伸表现方法的成熟；而这种独特的表现方法也正树立着 APScape 事务所的风格。

APScape 事务所的作品被用来进一步解释延伸表现是如何在设计项目中运用的。在事务所中，这一理念方法贯彻整个设计过程。它从对场地干预性的深入分析开始，通过速写、绘画、照片拼贴等，强调场地的特质和最重要的元素，同时研究干预的可能性以形成项目的主要想法。随着想法的发展，在上一步分析得出的场地逻辑被延续和充实，在最终的方案中能够看到它的影子。不同于常规的表现，最终的设计成果如剖面、效果、模型被诗意地进行表达，介于现实和抽象之间，展示部分而能“看”到全部，运用空白而引发想象……

The extended representation of the landscape is also a method that can be applied to the practice of design projects. Indeed, it is far more than a representation. In nature it is a design process. Valerio's representation and design are mutually influenced: the professional practice of his studio APScape nourishes the maturity and the methodology of the extended representation; and the unique representational method is shaping the APScape studio of its own style.

Works from APScape studio are employed to further explain how the extended representation can be realized in the design projects. The idea and method runs through the whole design process. It starts from a deep analysis of the intervention of the place through sketches, paintings, photo montages etc. which highlights the features and the most important elements of the place and at the same time it studies the possible interventions to form the main project idea. As the idea develops, the site logic, read through the previous phase, is continued and enriched, it is like a shadow that is perceivable in the final proposal. Differently from the normal representation, the final design products such as sections, renderings and models are expressed in a poetic way, between reality and abstraction, showing a part to "see" the totality, using the white to ignite imagination……

**红酒学校**

**意大利，阿韦利诺**

项目与周边环境：关于农业的绘画 The project and its context: drawing of agriculture

**Signs of landscape become architecture**

景观的符号成为建筑形式

**Harmony between architecture and landscape**
建筑与景观之间的和谐

**The architecture is part of the landscape**
建筑是景观的一部分

**The wine school and the relationship with the vegetation**
红酒学校与植物的关系

## 运动中心

**意大利，奥兰多总部**

Landscape - Voids
景观-空虚-填充

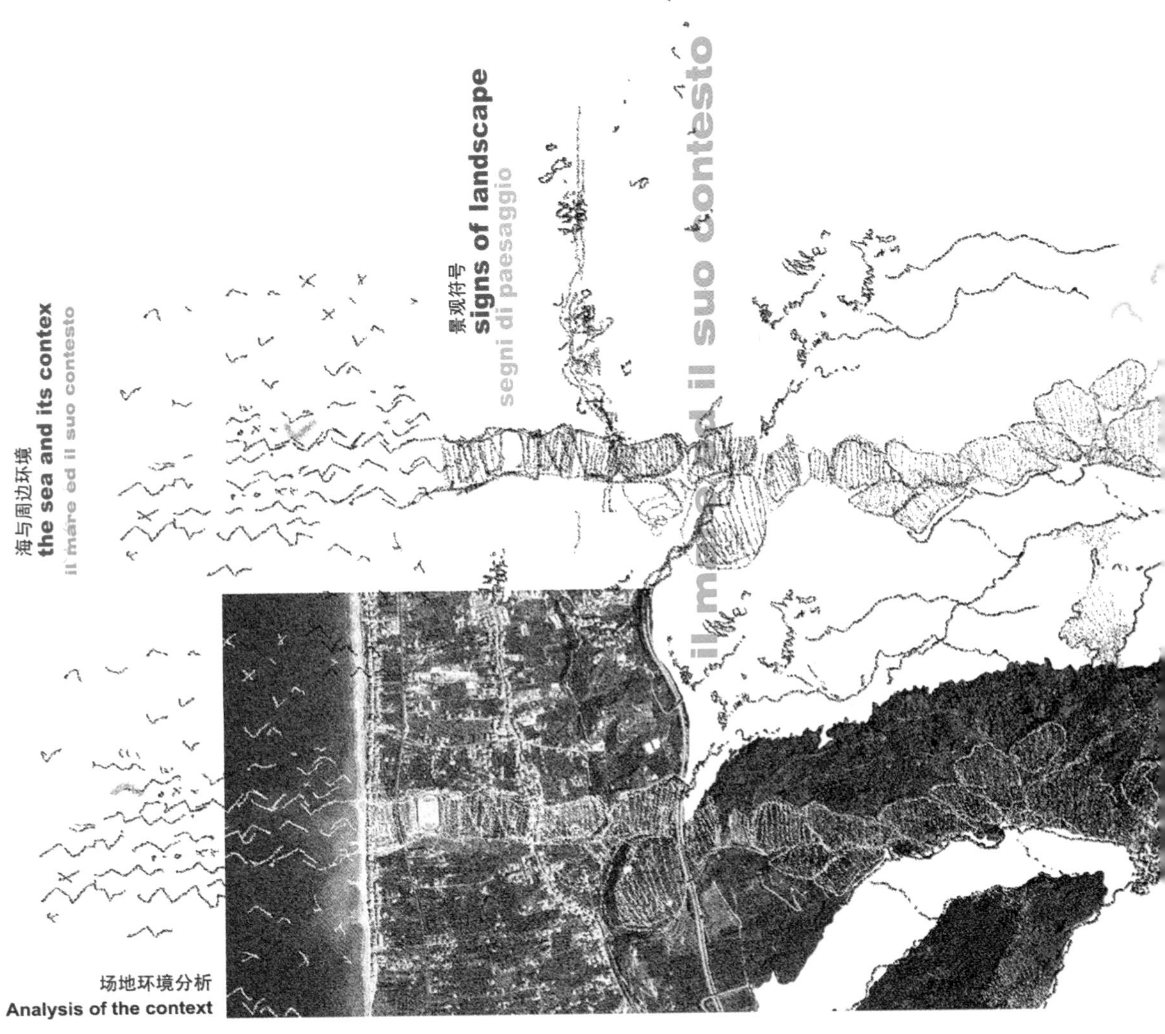


场地环境分析
**Analysis of the context**

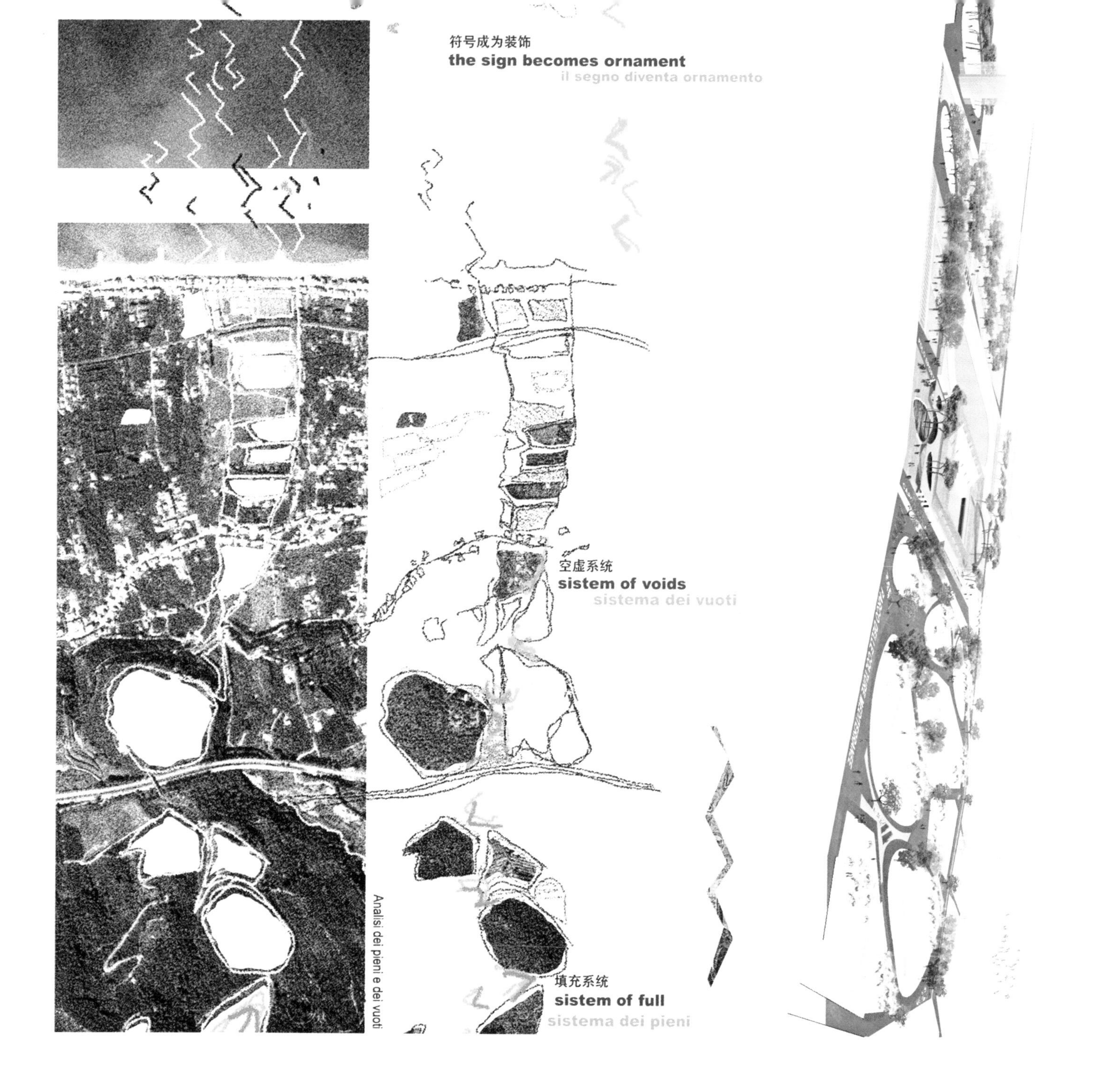
符号成为装饰
the sign becomes ornament
il segno diventa ornamento
空虚系统
sistem of voids
sistema dei vuoti
填充系统
sistem of full
sistema dei pieni
Analisi dei pieni e dei vuoti

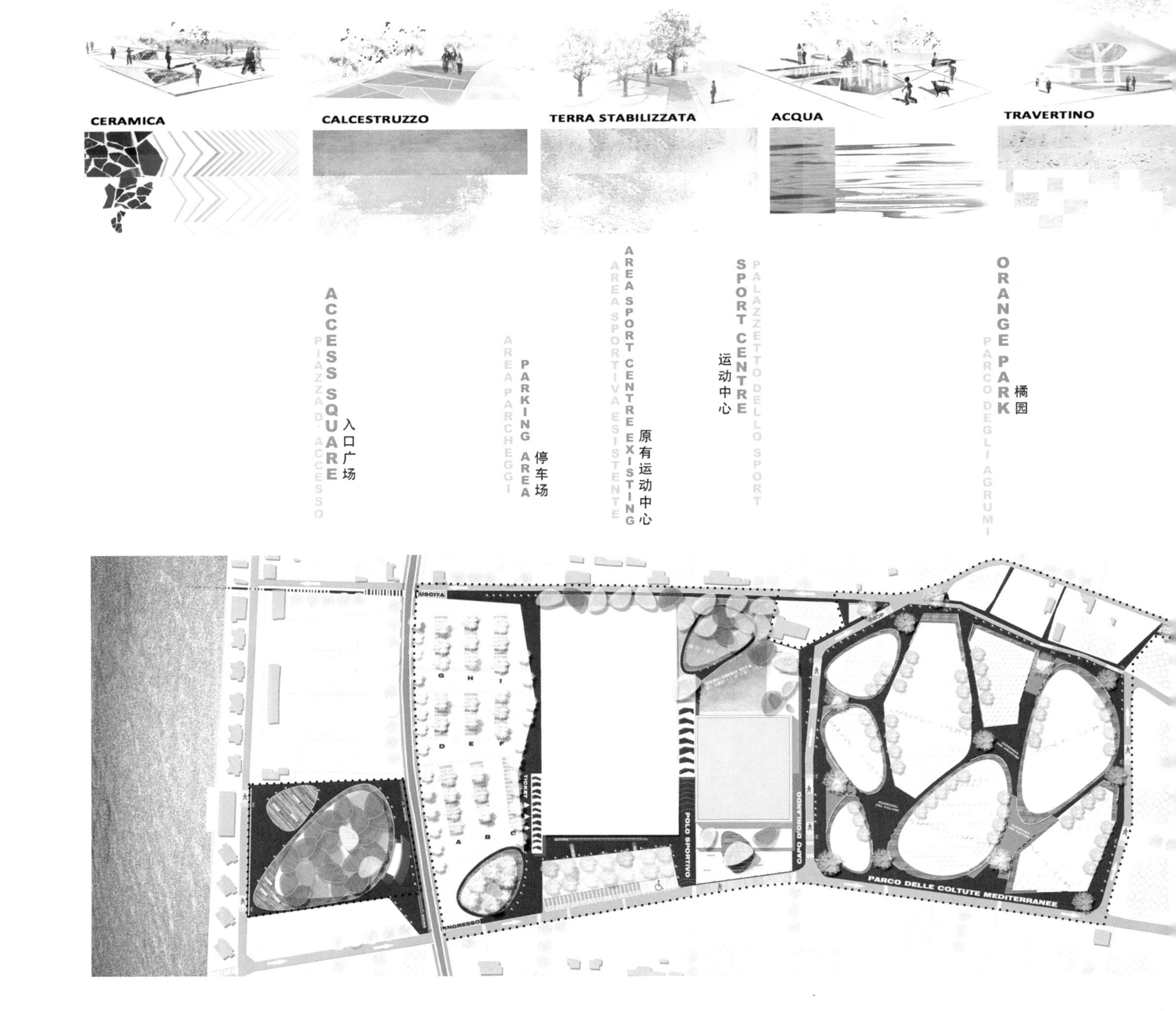

CERAMICA
CALCESTRUZZO
TERRA STABILIZZATA
ACQUA
TRAVERTINO
ACCESS SQUARE 入口广场
PIAZZA D'ACCESSO
PARKING AREA 停车场
AREA PARCHEGGI
AREA SPORT CENTRE EXISTING 原有运动中心
AREA SPORTIVA ESISTENTE
SPORT CENTRE 运动中心
PALAZZETTO DELLO SPORT
ORANGE PARK 橘园
PARCO DEGLI AGRUMI
USCITA
INGRESSO
POLO SPORTIVO
CAPO D'ORLANDO
PARCO DELLE COLTUTE MEDITERRANEE

ACCESS SQUARE
PARCO DEGLI AGRUMI
入口广场
SPORT CENTRE
运动中心
AVOLA
RAGUSA

鱼市

西班牙，韦尔瓦

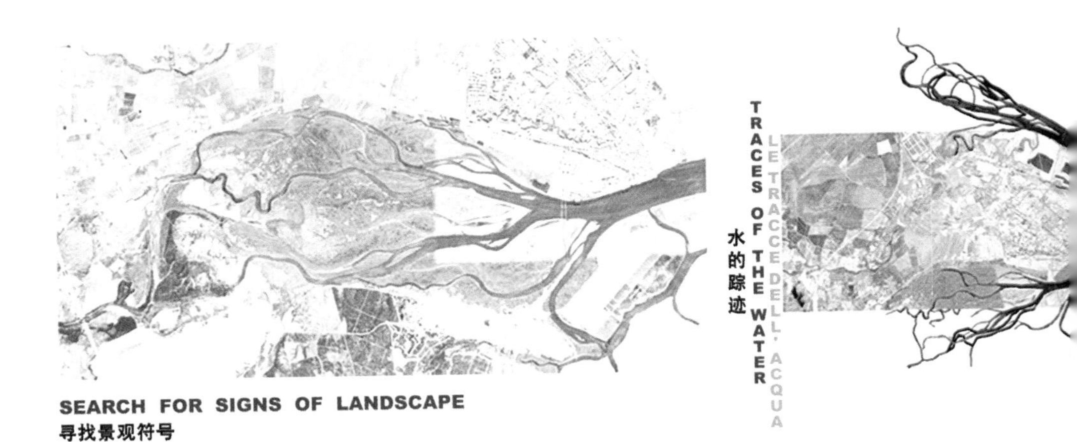

TRACES OF THE WATER
LE TRACCE DELL'ACQUA
水的踪迹


SEARCH FOR SIGNS OF LANDSCAPE

寻找景观符号

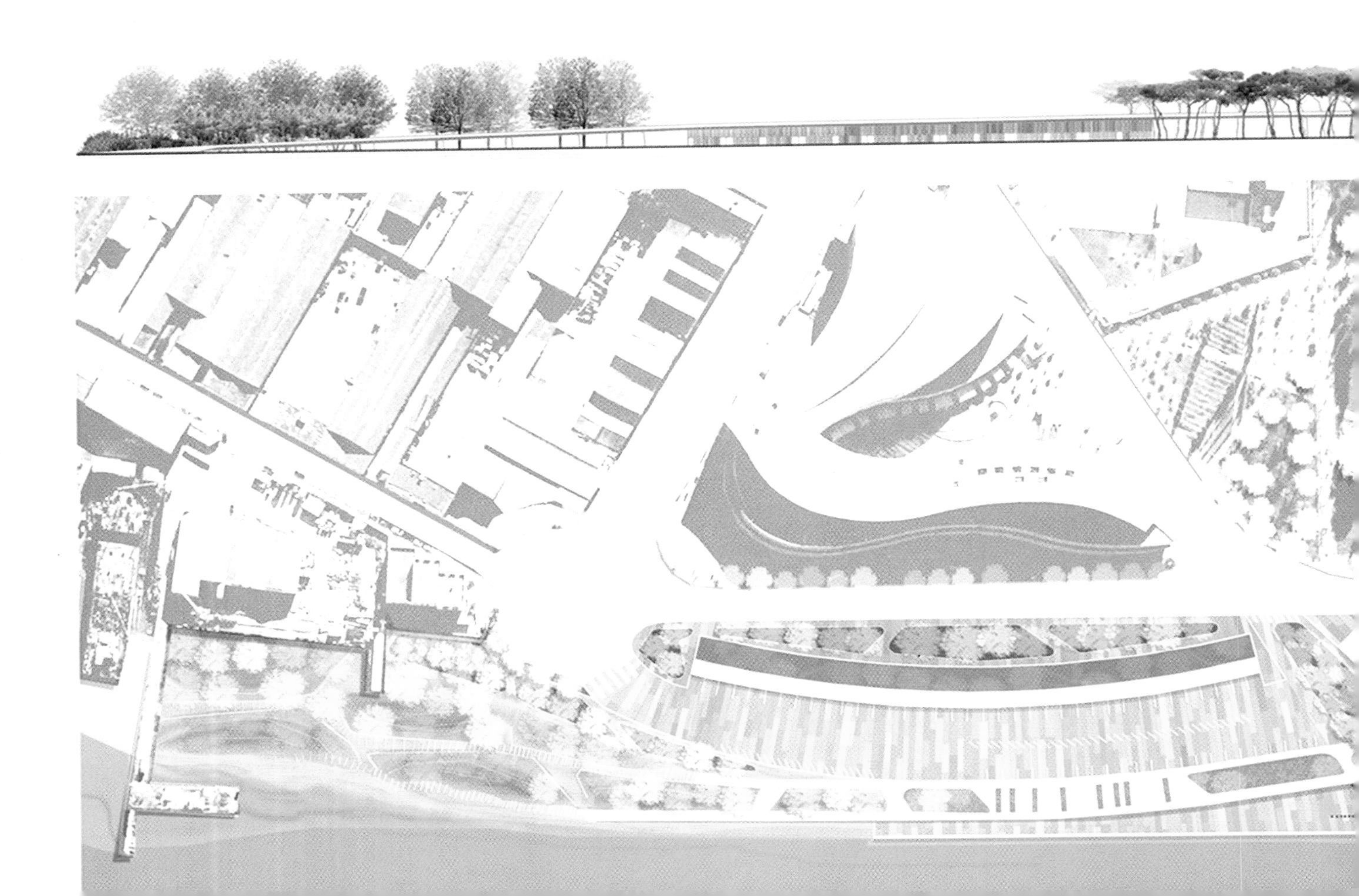

TRACES OF THE SOIL
LE TRACCE DELL SUOLO
土的踪迹

TRACES OF THE VEGETATION
LE TRACCE DELLA VEGETAZIONE
植被的踪迹

HYBRIDIZATION BETWEEN ARCHITECTURE AND LANDSCAPE
建筑与景观的混合

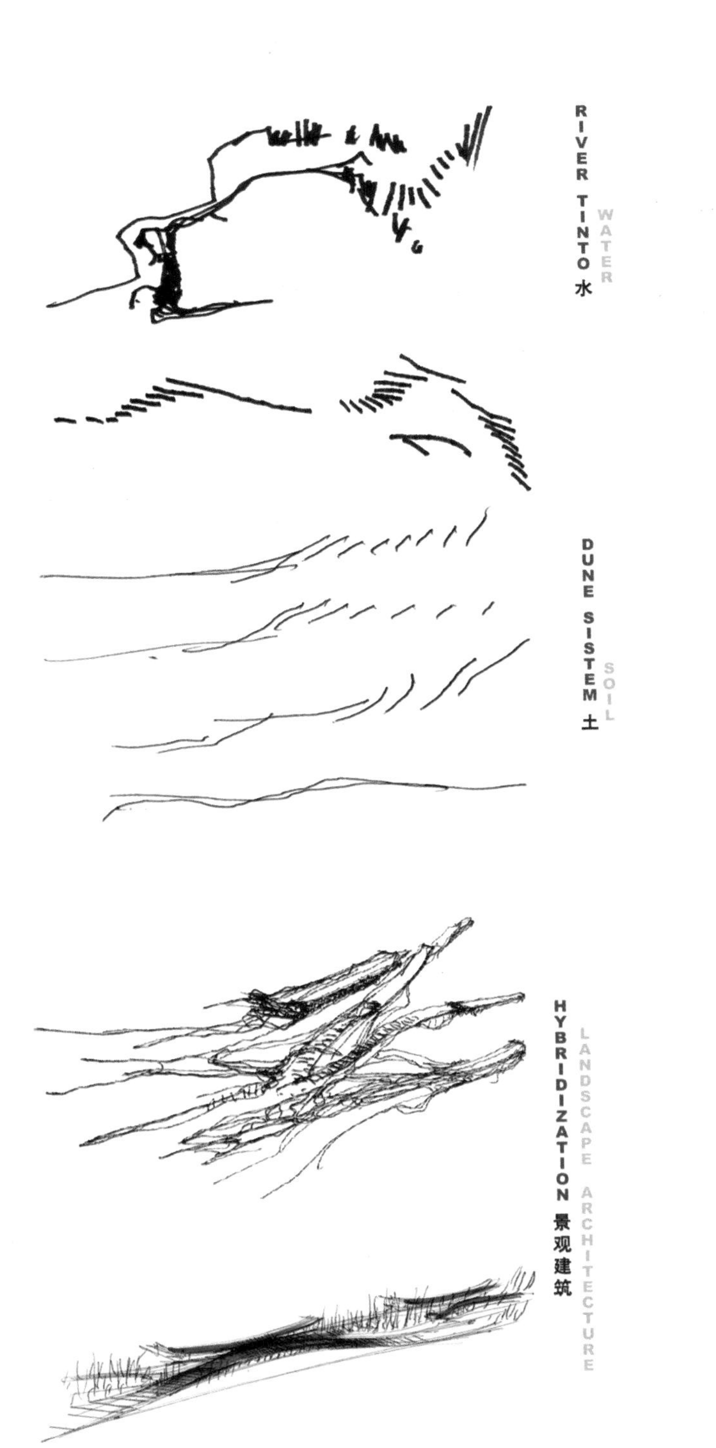
RIVER TINTO 水
WATER
DUNE SISTEM 土
SOIL
HYBRIDIZATION 景观建筑
LANDSCAPE ARCHITECTURE

LEVANTATO DEL SUELO
La cosa más abundante
sobre la tierra es el paisaje
Aunque todo el resto falta,
de paisajes siempre lo hemos
tenido de sobra
Hay épocas del año en que el
terreno es verde, otro amari
llo, luego marrón o negro
Y también rojo en ciertos
lugares pero éste depende de
lo que en el terreno se ha
plantado y se cultiva
o todavía no o no más
o bien de lo que os han
acido naturalmente
sin mano del hombre y
sólo llega a muerte
porque ha llegad
o su fin
No faltan los colores
a este paisaje
pero no sólo colores
y al mundo no faltan olores
José Saramago

# 别墅花园

## 意大利，西西里岛

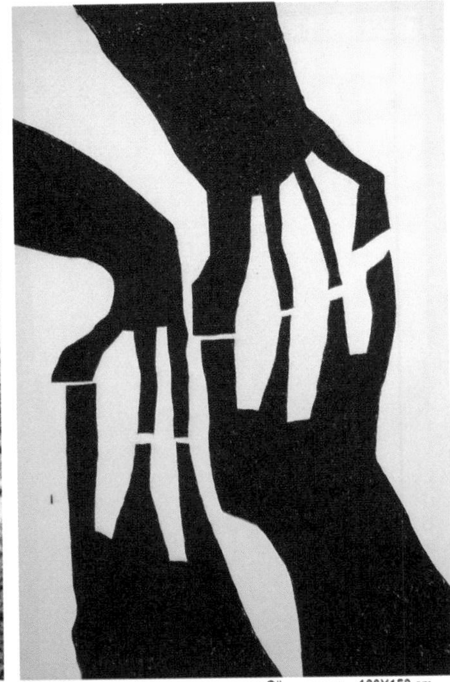

Oil on canvas - 100X150 cm

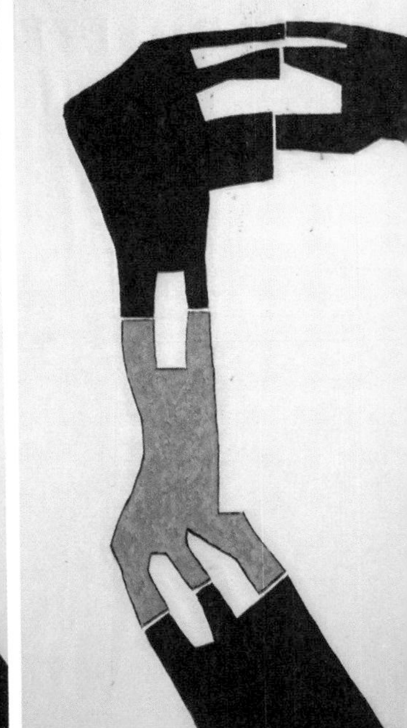

Acrylic on canvas

绘画-景观标志
**Drawings - landscape landmark**

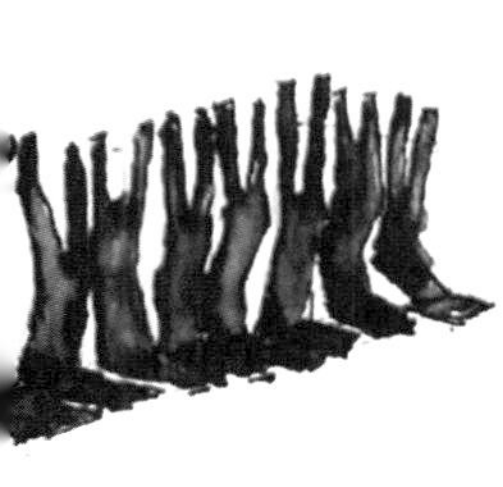

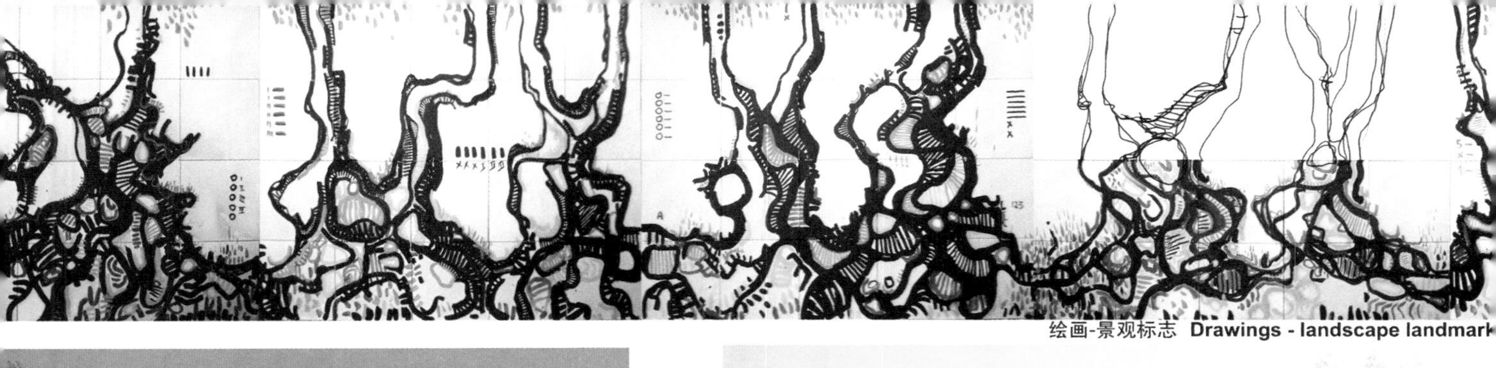

绘画-景观标志 **Drawings - landscape landmark**

Study of the elaboration of the landscape plan of the valley of the river Bouferkrane

河谷景观规划发展研究

# 景观规划

## 摩洛哥，梅克内斯

Masterplan and project phases 总体规划及分期

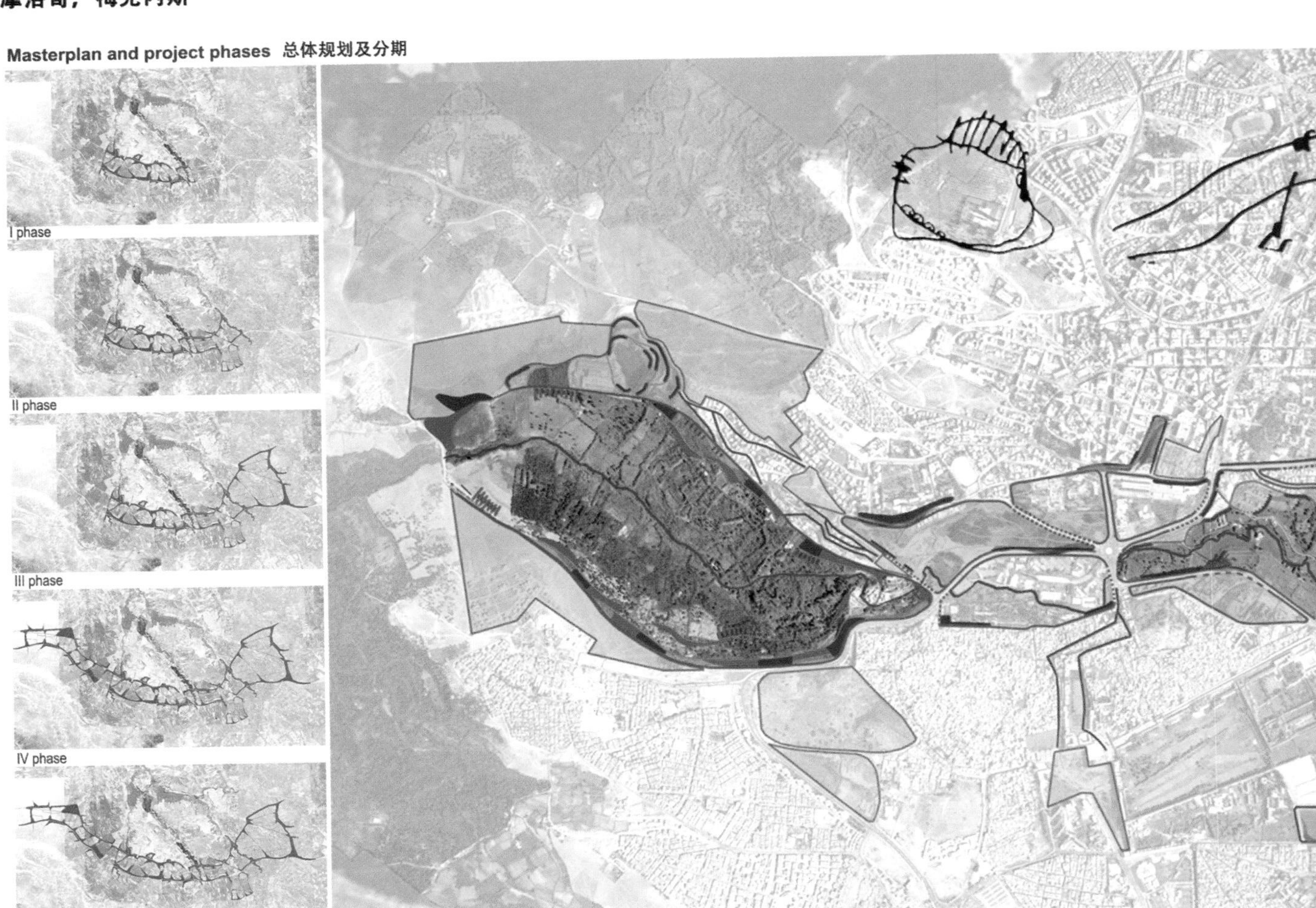

Recovery of green belts and agricultural system
绿带与农业系统的修复

Ecological park
生态公园

Agrucultural park
农业公园

Urban park
城市公园

Typological sections of paths of the park of the quarry
矿坑花园的路径剖面

Park of the quarry
矿坑花园

TOURISM OF IDENTITY
Mines of iglesiente
Sardinia - Italy


Drawing of mines
矿山图

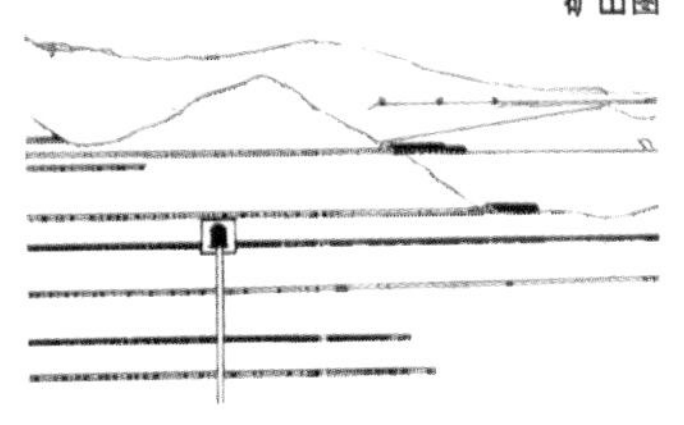

特色旅游
伊格莱希恩泰矿山
意大利，撒丁岛

矿山＋历史
**MINES** + HISTORY

植被＋多样性＋景观
VEGETATION + **DIVERSITY** + LANDSCAPE

符号＋特色＋形态
SIGNS + **IDENTITY** + MORPHOLOGY

景观＋生态＋系统
LANDSCAPE + **ECOLOGY** + SYSTEM

特色＋景观＋艺术
IDENTITY + LANDSCAPE + **ART**

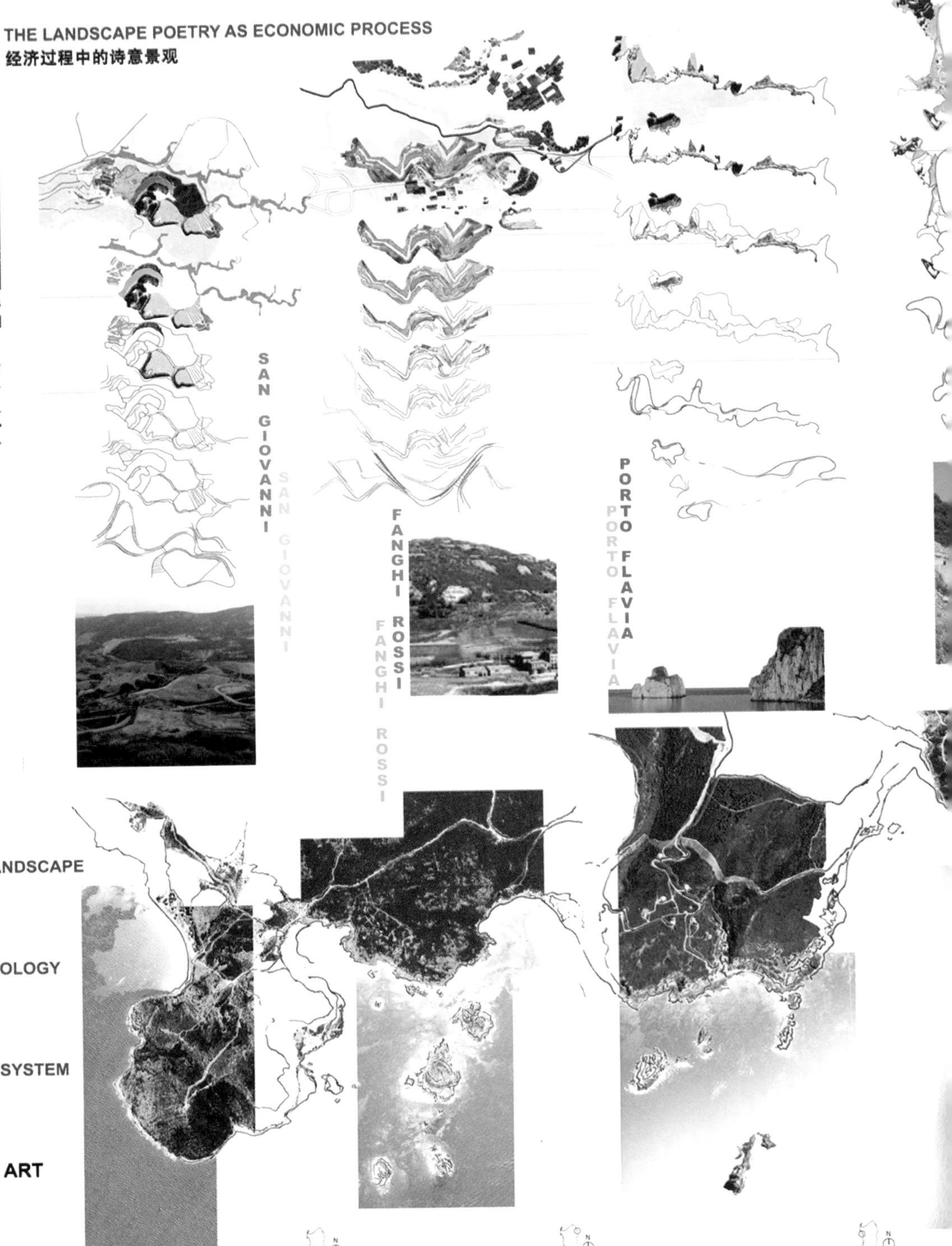

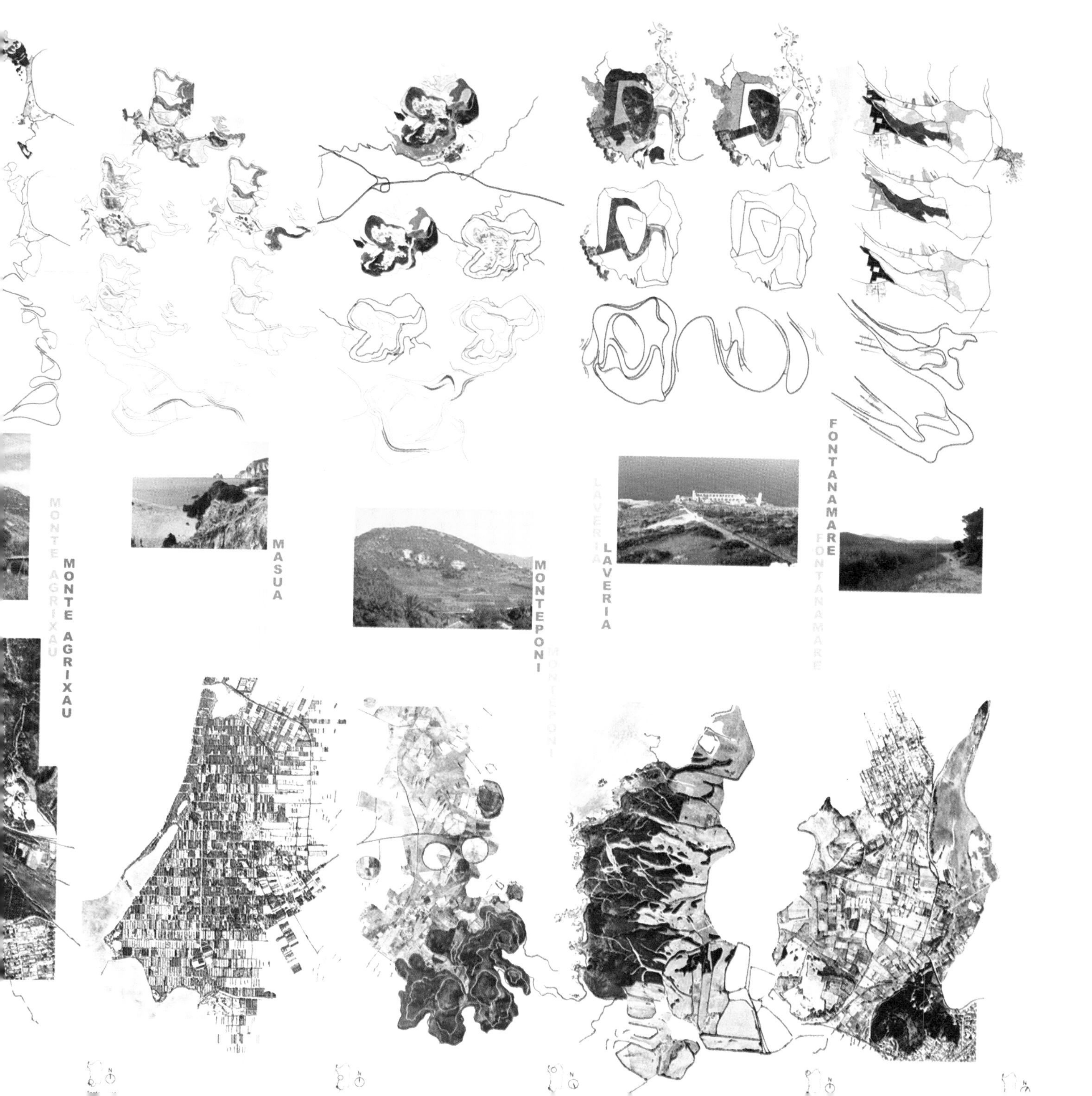
MONTE AGRIXAU
MONTE AGRIXAU
MASUA
MONTEPONI
MONTEPONI
LAVERIA
LAVERIA
FONTANAMARE
FONTANAMARE

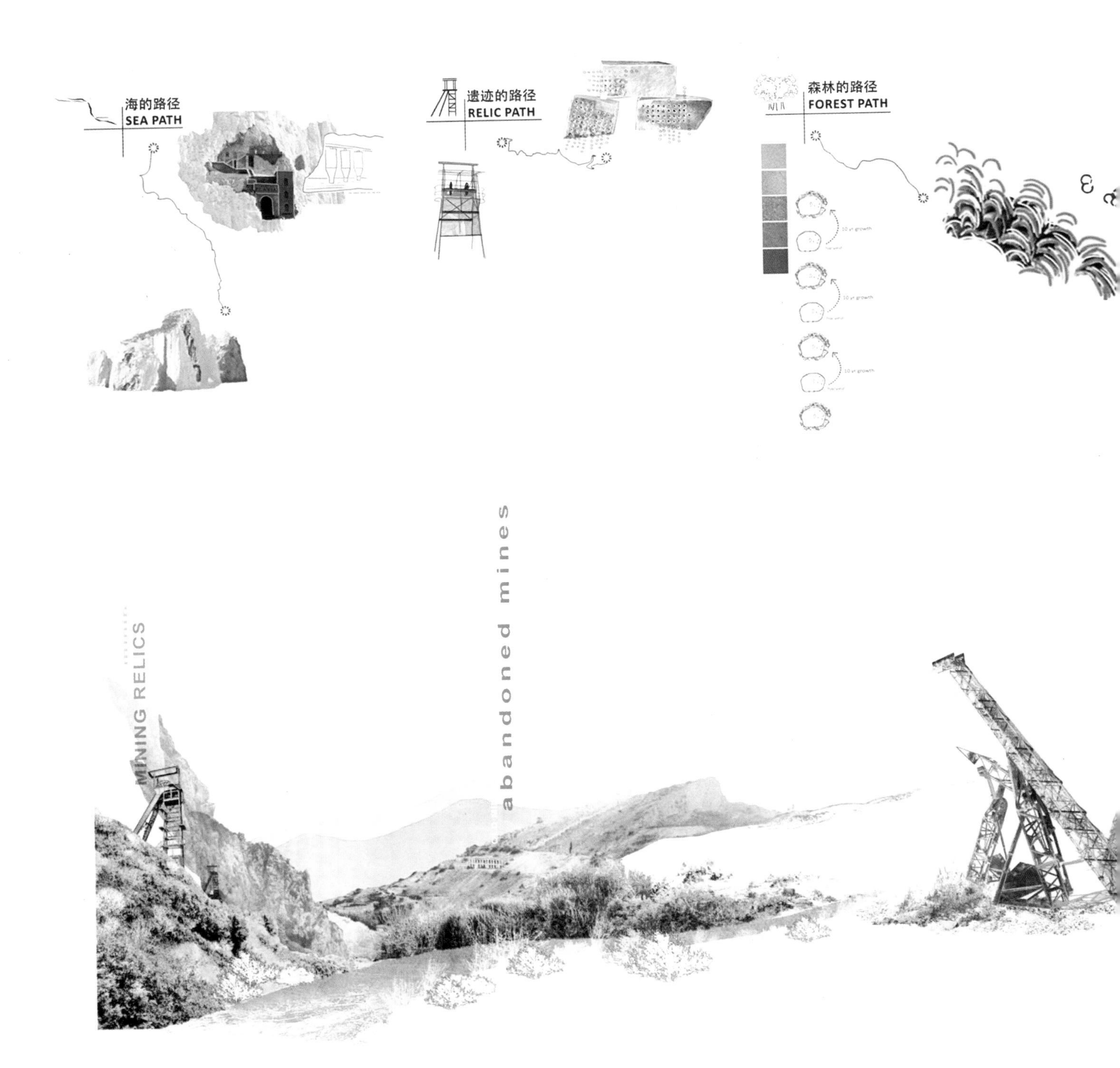

海的路径
SEA PATH
遗迹的路径
RELIC PATH
森林的路径
FOREST PATH
MINING RELICS
abandoned mines

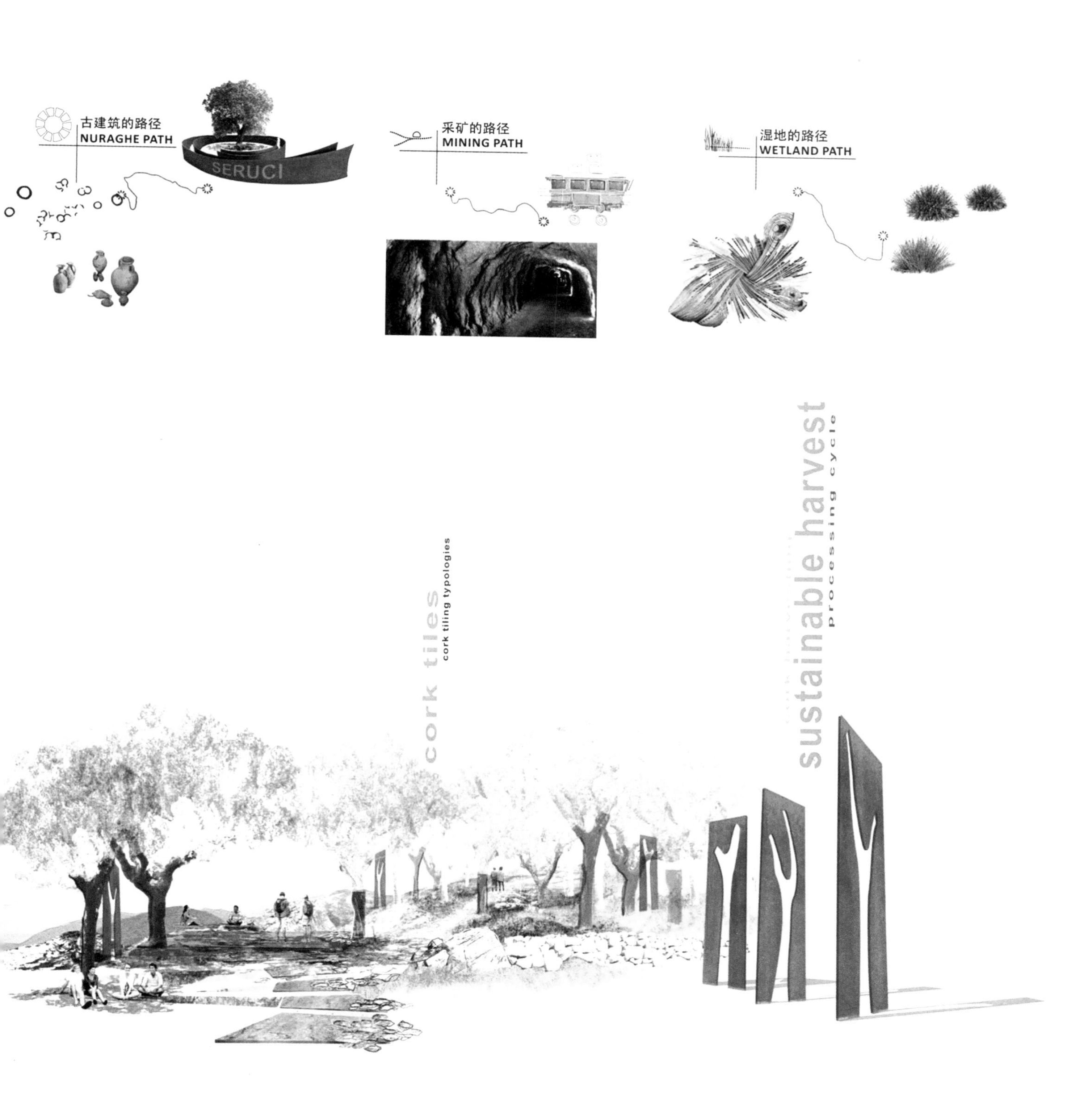
古建筑的路径
NURAGHE PATH
SERUCI
采矿的路径
MINING PATH
湿地的路径
WETLAND PATH
cork tiles
cork tiling typologies
sustainable harvest
processing cycle

MONTEPONIE RED MUD
MONTEPONIE
MONTEPONIE RED MUD
EDUCATIONAL PATH
biking
walking

San Giovanni mine
tunnels
4km MONTE SAN GIOVANNI

合悦百翠农庄概念规划

中国，浙江桐乡

公园+度假+有机农场

PARK+RESORT+ORGANIC FARM

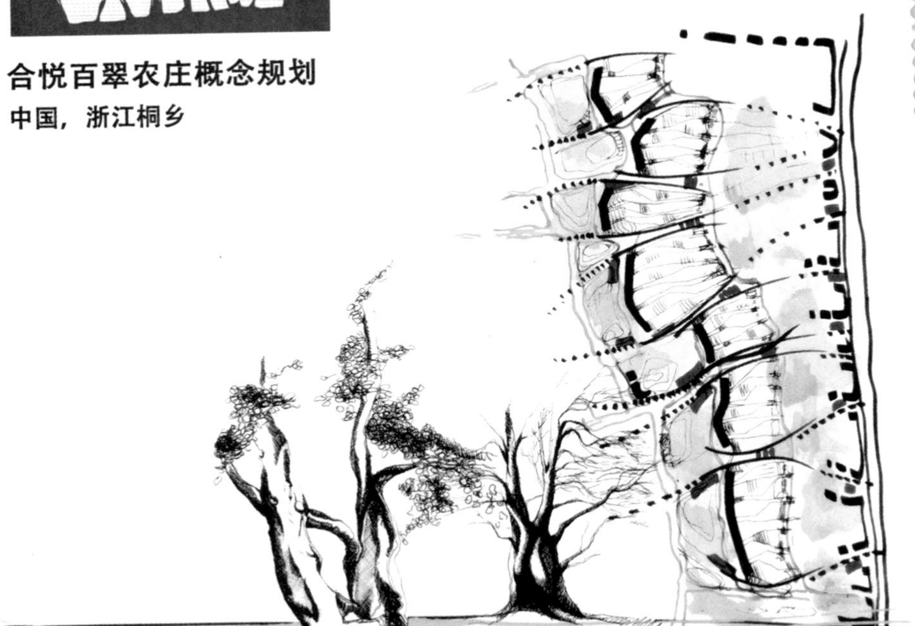

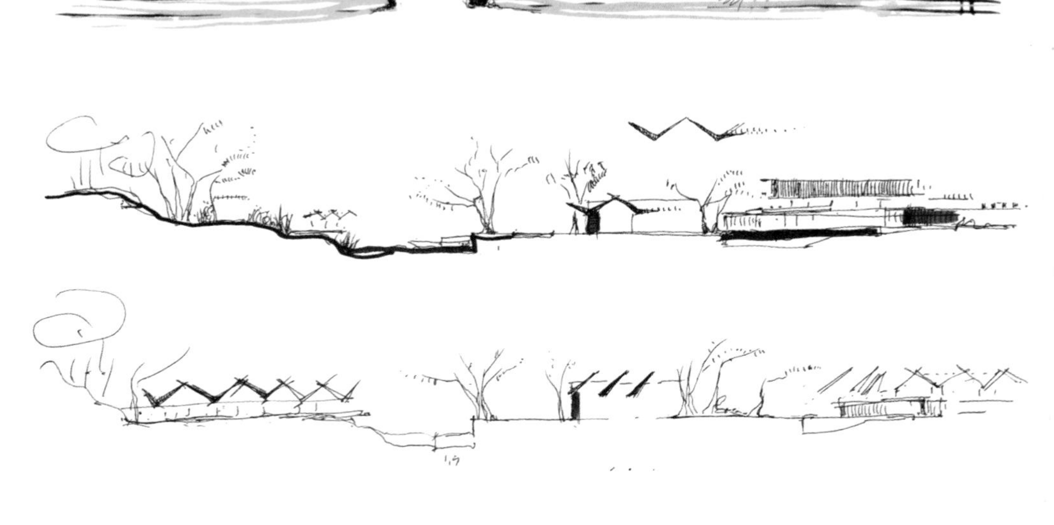

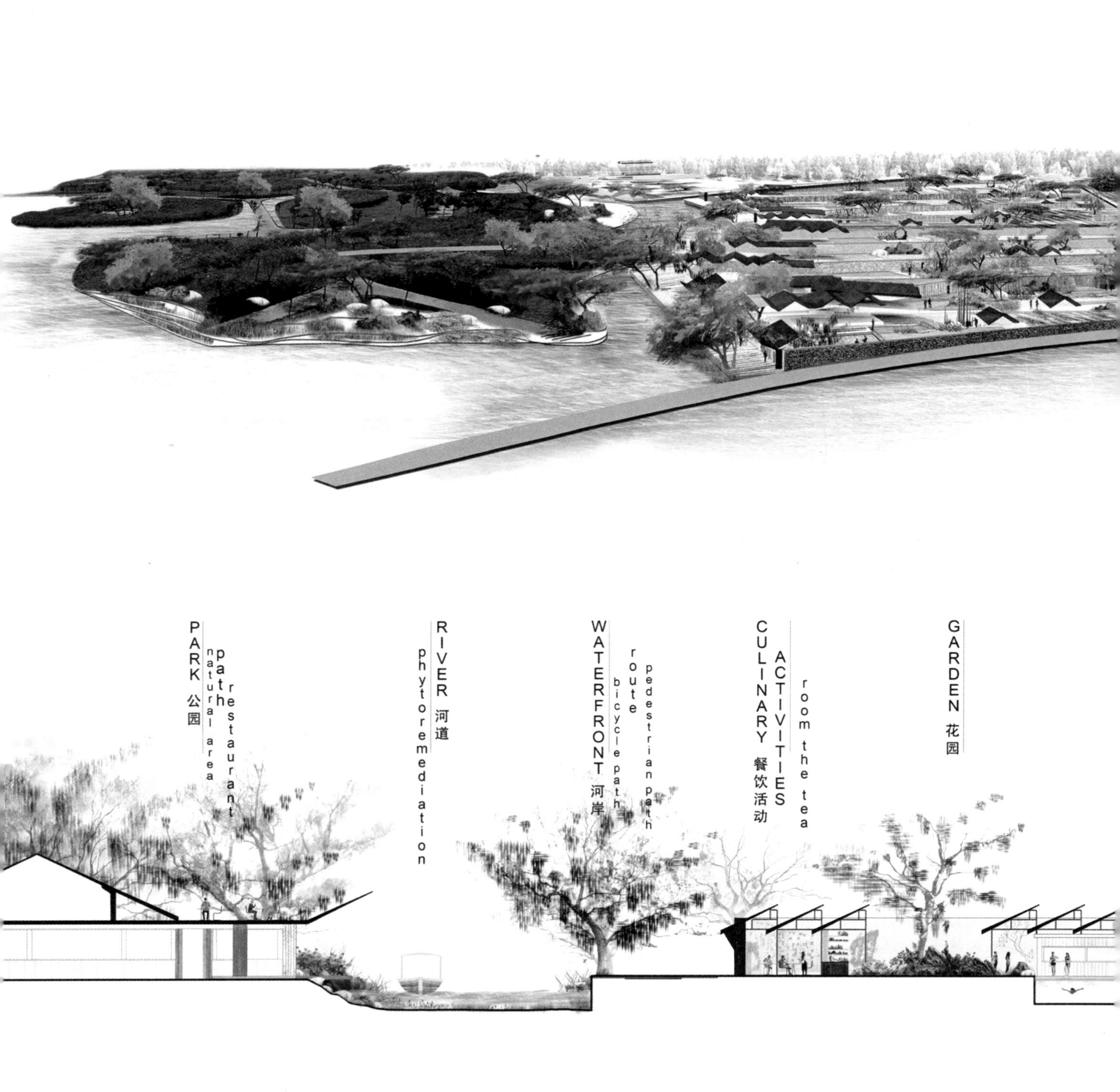

PARK 公园
natural area
path
restaurant
RIVER 河道
phytoremediation
WATERFRONT 河岸
bicycle path
route
pedestrian path
CULINARY 餐饮活动
ACTIVITIES
room the tea
GARDEN 花园

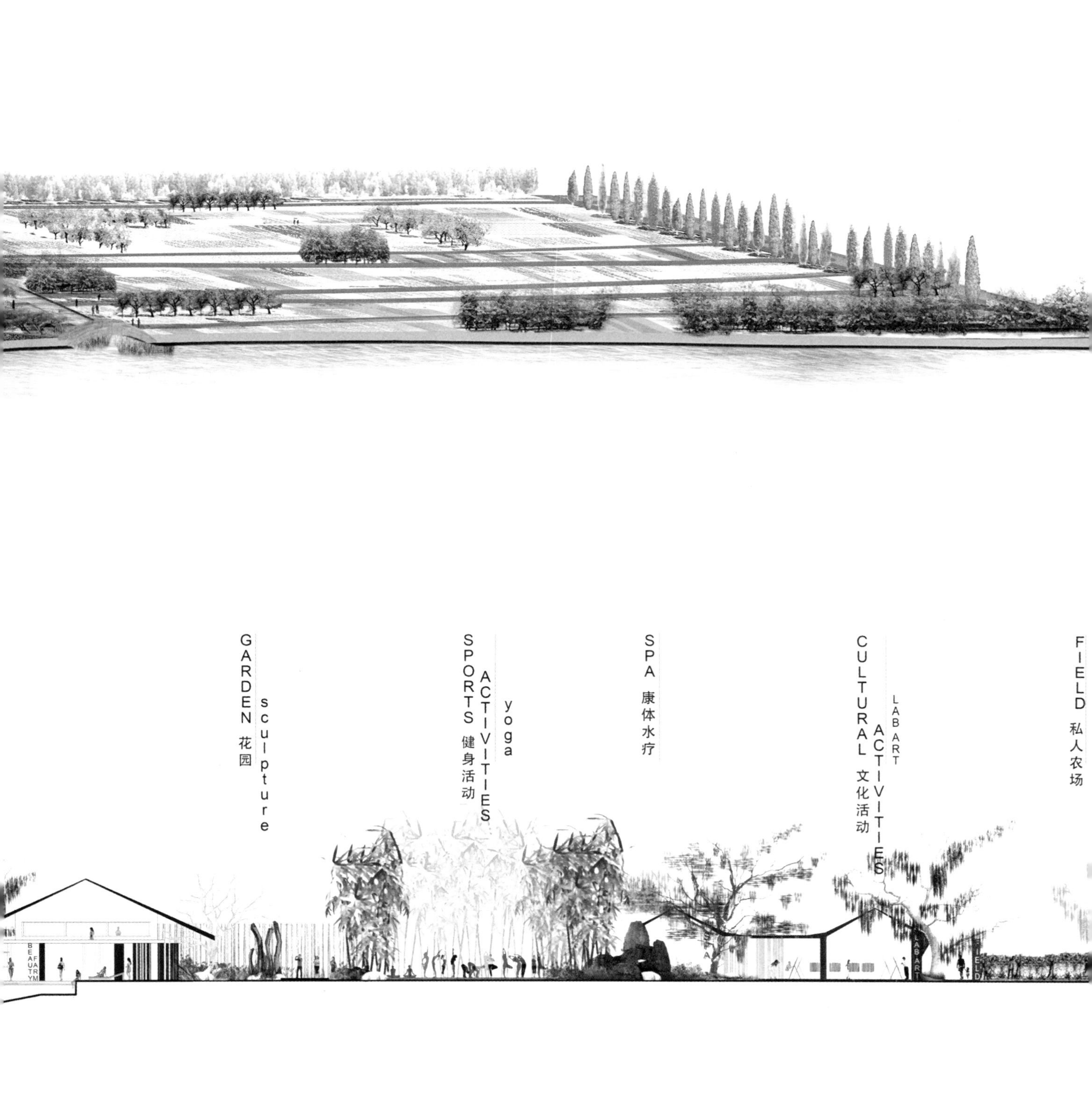

GARDEN 花园
sculpture
SPORTS 健身活动
ACTIVITIES
yoga
SPA 康体水疗
CULTURAL 文化活动
ACTIVITIES
LAB ART
FIELD 私人农场

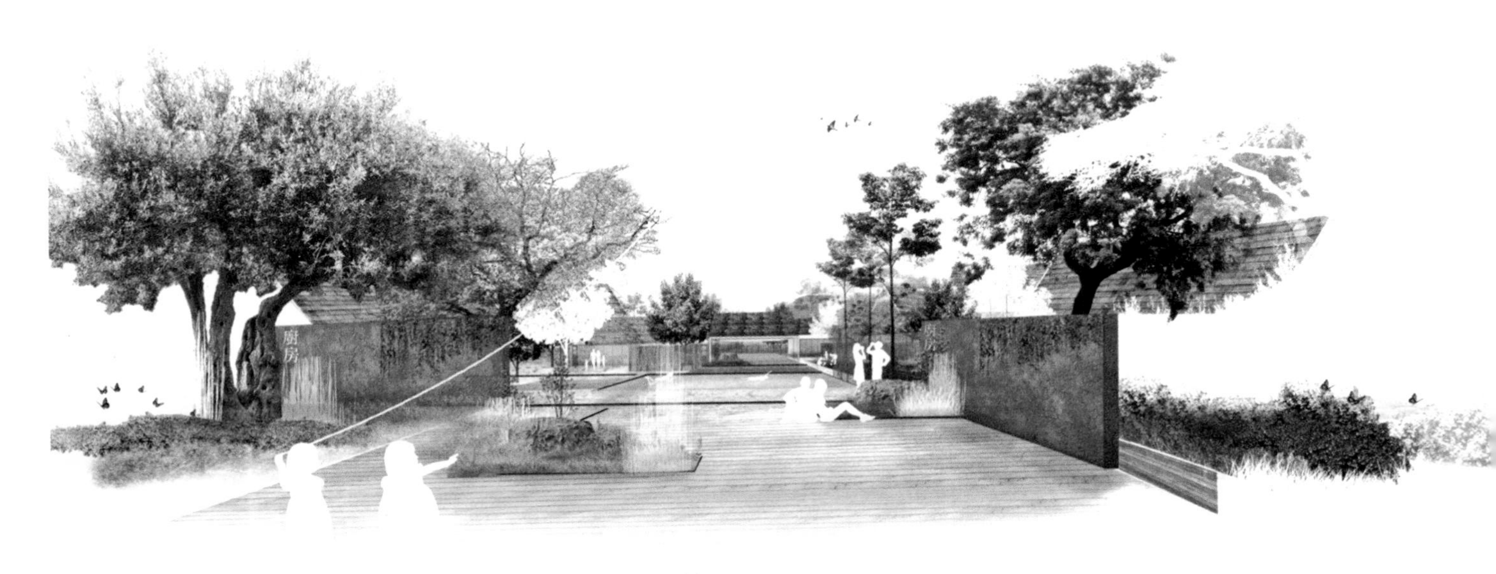

**Integration of walls, buildings, courtyards, terraces, gardens and farms**
墙体，建筑，庭院，露台，花园和农场的有机整合

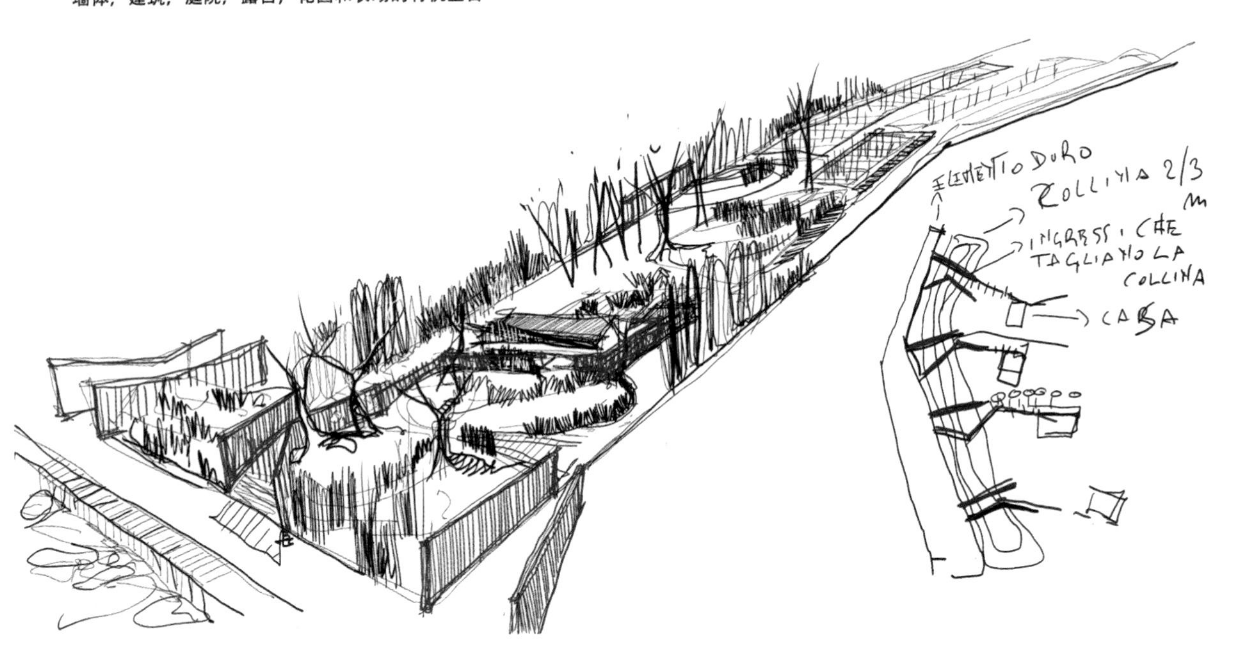


**Spatial sequence of the Resort Island**
度假岛空间序列

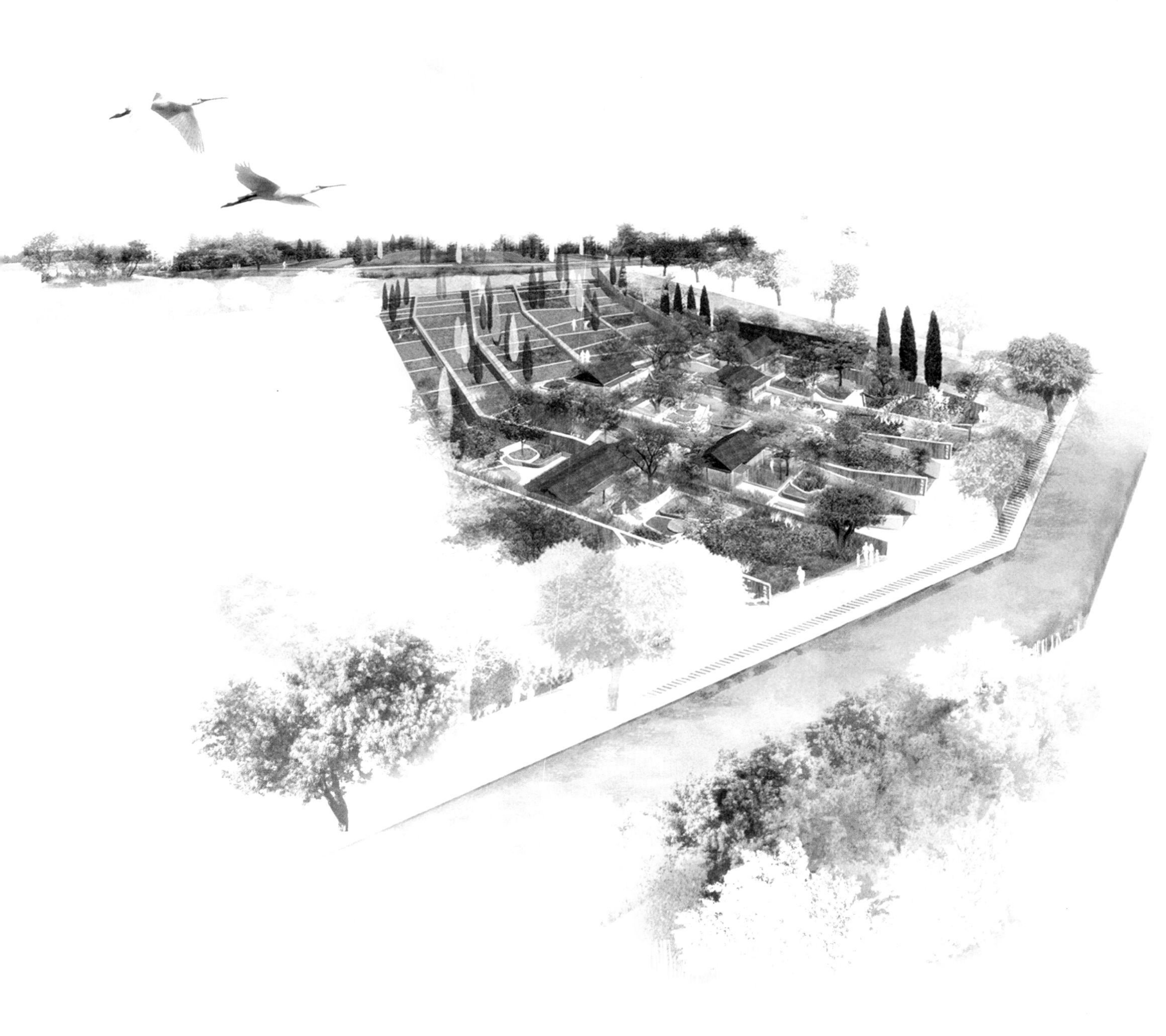

农业为主题的度假岛　**Agriculture-themed Resort Island**

农业展览温室和水生作物展区
Green House for agriculture exhibition / Water plants exhibition

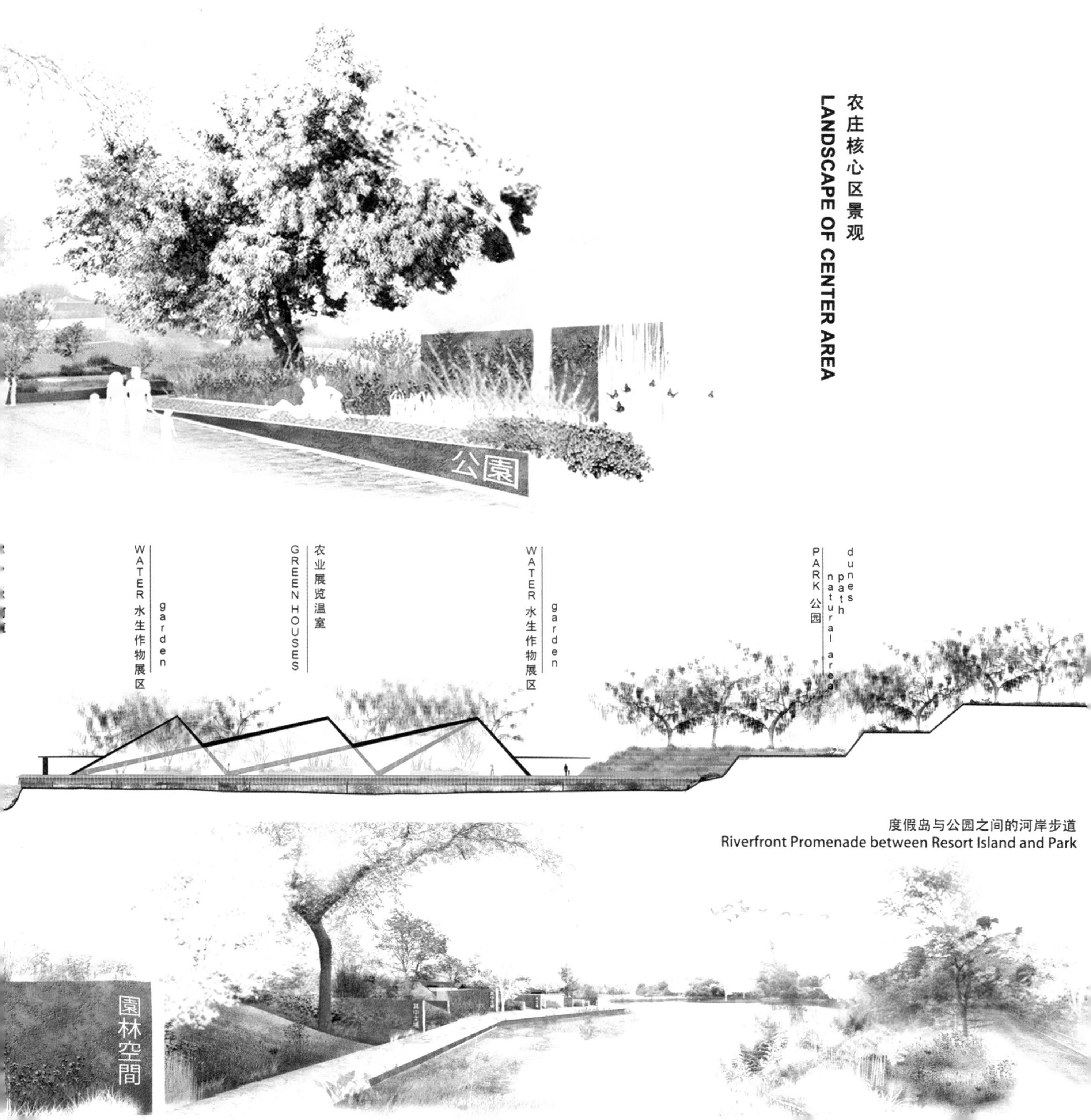


度假岛与公园之间的河岸步道
Riverfront Promenade between Resort Island and Park

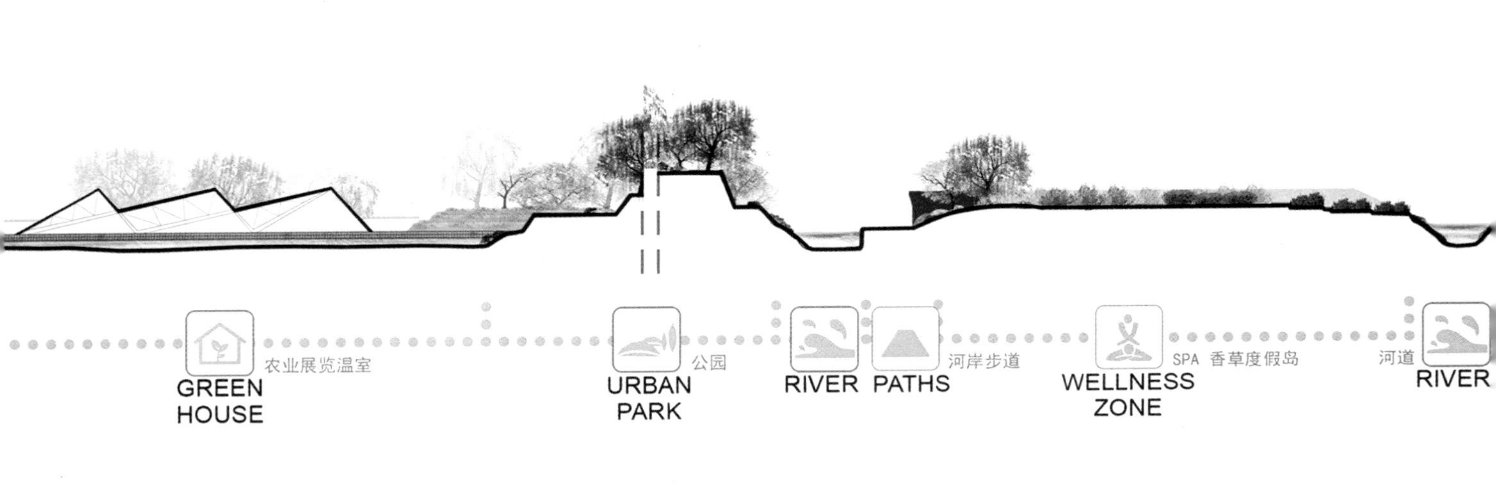


农业展览温室和水生作物展区
Green House for agriculture exhibition / Water plants exhibition

停车场
RE PARKING
Landscape Island
景观岛
STESSA ALTEZZA
LA PERGOLA ENTRA DENT
PIU' DISTESE

**Garden and farm on Resort Island** 度假岛花园农场

Oganic Farm 有机农场

以本土果木和艺术装置为特色的系列林荫道

**Series of parkways featured with local fruit trees and artworks**

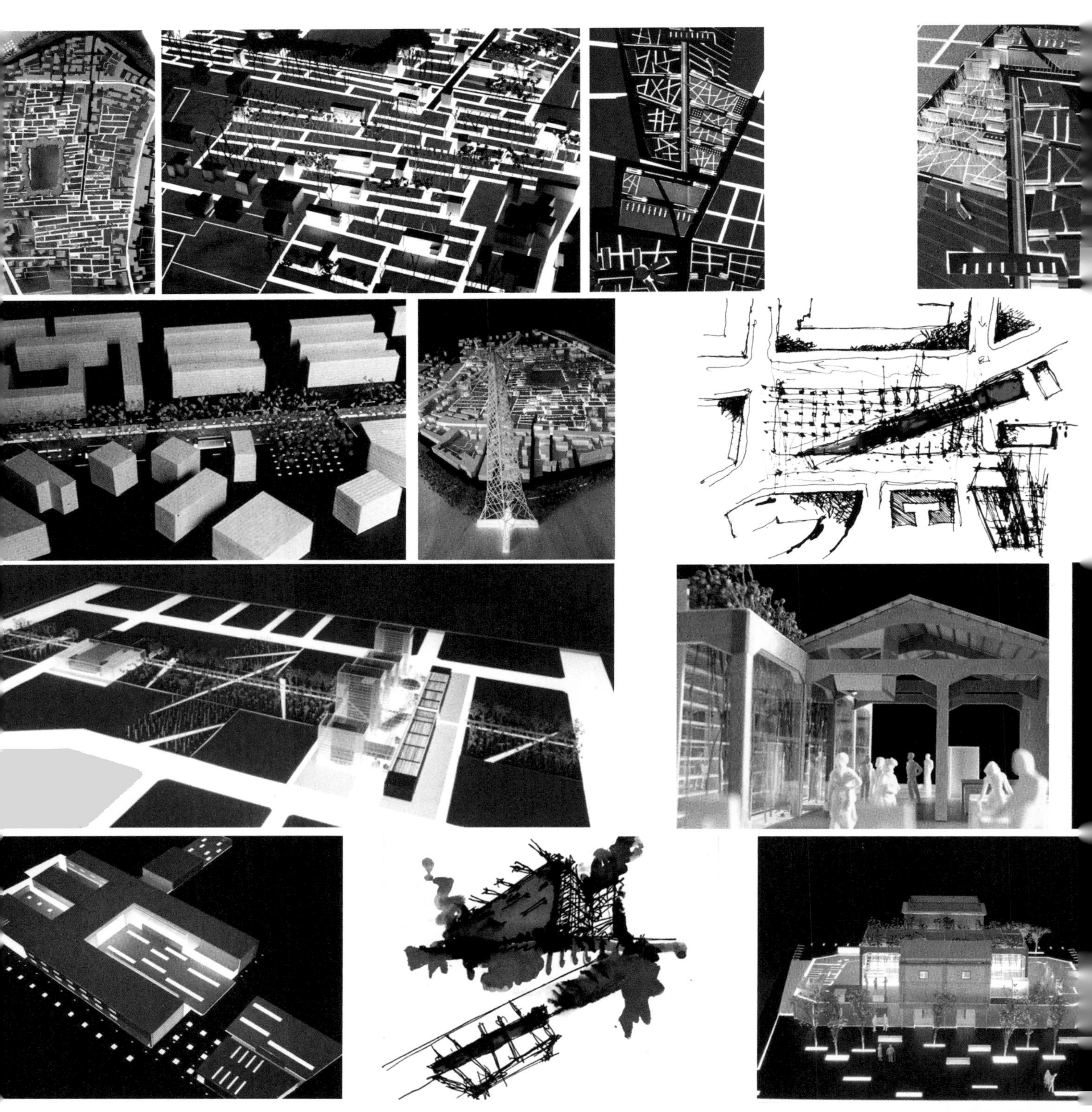

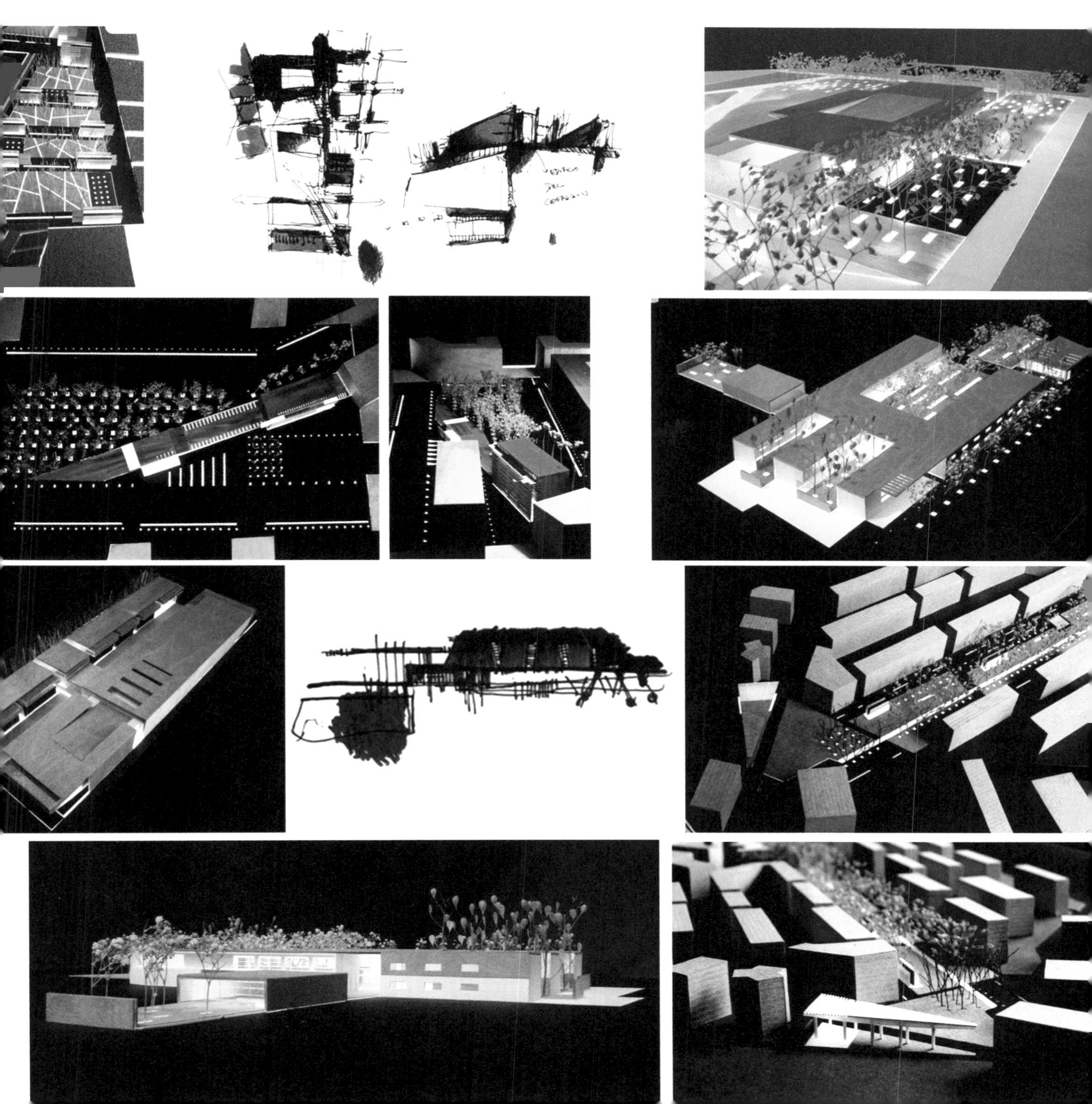

# 致谢 Acknowledgements

从写这本书的最初念头到她的出版，集聚了来自众多可爱人士的盛情帮助。首先要感谢我们最敬爱的导师大卫·加弗努尔，系主任詹姆斯·科纳和理查德·韦勒，还有系主任助理黛安娜·普林格女士，感谢他们邀请我们去宾夕法尼亚大学景观系访学，并给予我们如此美妙的学习和研究经历。然后，我们要对瓦莱里奥·莫拉比托教授说无数遍的感谢，为他邀请我们参加他发人深省的课程及其无比精彩的期末评审，还有他赋予我们写此书的机会和支持。也感谢宾大亲爱的学生们，为他们对创造性想法和优秀作品的分享。此外，感谢瓦莱里奥教授工作室的所有成员，他们提供了精美的设计项目图纸及成果，尤其是斯蒂芬尼亚·孔杜尔索，她帮忙做了大量选择和组织工作。还要感谢同济大学景观学系教授及博士生导师金云峰对本书的启发性建议，以及刘颂教授和韩锋教授的支持。最后，感谢索尼亚·比拉尔迪、王南、陈融，以及出版社的编辑徐纺、滕云飞和孙冰若对文字的校对和编辑。

项淑萍

弗朗切斯科·贝利杰兰特

From the first idea of having this book until its publication, it has accumulated a lot of great helps from many lovely persons. First of all, thanks to our dearest supervisor David Gouverneur, Chairs James Corner and Richard Weller, the Assistant to the Chair Ms. Diane Pringle, for inviting us to the Department of Landscape Architecture at University of Pennsylvania, as visiting scholars, to have such great study and research experiences. Then, we have to give endless gratitude to Professor Valerio Morabito, for inviting us to his inspiring class and the fantastic final review, giving us the opportunity and support to write this book. Thanks also to the dear students, for their generous sharing of creative thoughts and amazing works (see students' name). Besides, thanks to Valerio's studio (see members' name) to provide their beautiful drawings and products of design projects, special thanks to Stefania Condurso for her helps to choose and organize their works.In addition, many thanks to Professor and PhD Supervisor Yunfeng Jin, in the Department of Landscape Studies at Tongji University, for his enlightening suggestions for the book, also thanks to Professor Song Liu and Professor Feng Han for their support. Finally, thanks to Sonia Bilardi, Nan Wang, Rong Chen and editors of the publisher Fang Xu, Yunfei Teng and Bingruo Sun, for helping to check and edit the text.

Shuping Xiang

Francesco Belligerante

## University of Pennsylvania

LARP 720- Fall 2012:
Burrows Susanna B.
Chen Fei
Cua Albert C.
Doherty Barrett H.
Fang Chenlu
Han,Sa Min
Jo Yong Jun
Kaplan Taylor A.
Lee Chi-Yin
Lee Jeong Hwa
Lee Jiae
Lee Kyung Keun
Li Qiyao
Li Wen
Li Yiran
Lu Wenwen
O'Connor Kyle D.
Tao David
Tian Siyu
Viquez Dana L.
Wang Jiaqi
Yin Mingyu
Yoon Eugenia S
Zhao Yitian
Zhu Jianchun

LARP 720 - Fall 2011:
Chen Wei
Chiarelli Elizabeth L.
Cho Koung Jin
Choi Minyoung
Farquhar Kathryn H.
Gupta Ekta
Jackson Justin D.
Jankowsky Margaret
Kwon Yu
Lee Sanghoon
Li Chunjin
Liu Hanxiao
Loh Andrew W.
Ludwig Ashley B.
Mcconnico Stephen A.
Mcelroy Nicola M.
Nam Hyunjoo
Rufe Kathryn M.
Schneider Ann M.
Schwartz Sandra C.
Shi Xiayao
Shin Da Young
Song Ting
Storm Meghan T.
Tsay Meng-Lin
Tse Hong Cheng
Wolf Sarah D.
Zhang Rui
Zhang Yuanling

LARP 720 - Fall 2010:
Barthmaier Johanna F.
Boland Rana J.
Burgi Stephan L.
Carter Leslie J.
Chang Po-Shan
Chiu Yi-Ting
Degregorio Michael A.
Hart Marie F.
Henry Tamara M.
Huang Kerry W.
Kim Caroline S.
Kim Hyun Suk
Kwan Wing L.
Lee Ho-Young
Lim Sanghyun
Lin Connie P.
Linsenmayer Amy A.
Liu Sheng
Lutsky Karen O.
Marwil Joseph I.
Moin Sahar
Park Soohyun A.
Rabiee Zeinab
Ragulina Svetlana
Shim Bowon
Tsutsumi Yuichiro
Vazquez Alejandro D.

## APS Spinoff UNIRC

Valerio Morabito
Debora Gallina
Martino Milardi
Gianpiero Donin
Stefania Condurso
Alessia Latella
Daniele Saporita
Daniele Politi
Carmine Carfa
Rachele Sergi
Università degli Studi di
Reggio Calabria
Tecno Habitat

Main Partners:
Iman Benkirane
Ugo Sgambetterra
Sebastiano Micali
Antonio Gabbellini

Teaching Assistant:
Megan Burke
Karen Lutsky
Nicholas Pevzner

(Cooprogetti cooperated in the projects: School of Wine, Waterfront Pantelleria)

# 后记

电影《里斯本的故事》中有这样一幕场景，片中的导演对影像麻木了，于是将摄像机搁在背上，用这种“反”的方式来记录不经过他眼睛观察的、真实的里斯本。从这一有趣的故事联想到自己平时“只顾眼前”的空间阅读方式，恍然发现一丝局限性。后来，瓦莱里奥教授讲述了伊塔洛·卡尔维诺“源自不透明性”的小故事，那时对空间阅读彻底觉悟了一番：每个人都被三个无限的维度穿越，前面后面上面下面左面右面，随着身体的移动，这些维度动态转换着……将之结合到里斯本的故事，体验空间时我们的身体必须向六个方向张开，就像安上了六个隐形的摄像机，摄取每个维度上无限延伸的信息，这样才可能获得空间的本质。

表现空间的观念也因此需要改变。对于传统的表现课来说，瓦莱里奥的表现理念简直就是彻头彻尾的反例。“不要画你所见”、“画错误”、“画风，不要画树”……这些就是瓦莱里奥表现课上的经典语汇。经历了革新式的洗礼后，我们的那些老观念被悄无声息地淹没了。他对表现技术放宽到了几乎没有底线，从来不评判你画得美还是丑，让人觉得你应该会画，很会画。他引导的重点是画什么？为什么要画？为什么要去表现？走进空间，走进图片，获取六个维度上的信息，甚至延伸到你的记忆与知识中的信息，把你对空间或图片的独到理解，想方设法表现到极致，将个性放到最大。简言之，就是要画“心”，画“想法”。透过画，可以看到一片更大的图景；透过你，可以发现一个另类的世界。

两个有趣的故事以及瓦莱里奥颠覆式的表现思想与方法，催生了写这本书的念头。撇开无穷无尽的表现花样和技巧，简单回归到自己的想法，回归到空间的本质，未免不是一种更深刻的尝试。很有意思的是，与瓦莱里奥的教学探讨中，我们发现了他的延伸表现方法与中国国画的相通性，即诗意。通过这本书的阅读，相信不难发现。所以，永远都不要忽视和放弃你垂手可得的宝藏。

项淑萍<br>弗朗切斯科·贝利杰兰特

# From the Authors

In the movie Lisbon Story, there is one scene in which the director is overwhelmed with images; instead of shooting with his eyes, he puts the camera on his back, trying to film the true Lisbon in such a reversing way. This interesting story reminds us of our ordinary way of reading the space, which is only through the eyes; and then suddenly, we realized the limitation. Afterwards, Professor Valerio told us the story of Italo Calvino's Dall' opaco: the conception that each of us is crossed by three infinite dimensions, in front, behind, above, below, to the right, and to the left, as the body moves, these dimensions change dynamically…… At that time, we got a thorough enlightenment of our ways of reading the space. If integrated with Lisbon Story, we should open the senses of our body in six directions, as six invisible installed cameras, in order to get the infinite information from each direction. Then, we may capture the essence of the space.

Accordingly, the idea of representing the space needs to be changed. Compared to the conventional representation class, Valerio's idea of representation is a complete subversion. 'Don't draw what you saw', 'Draw the mistakes', 'Draw the wind, not the trees'... These are the typical sayings in Velario's representation class. Going through a revolutionary mind changing, our conservative ideas were fading away. For Valerio, there is no bottom line in the representation technique, and he will never judge whether your drawing is beautiful or ugly, instead, he makes you feel confident on drawing, in a very joyful way. The focus of his teaching is: What to draw? Why to draw? Why do we need to represent? Go inside the space, go inside the image, get the information from six dimensions, or even extend it to your memory and knowledge, express your unique understanding and feeling of the space or images, represent and invent whatever you can, amplify your unique personal idea. Briefly, it is all about drawing your "idea" with your "heart". Through the drawing, people are brought into a bigger picture; through you, people get into a special world.

These two interesting stories and Valerio's innovative representation idea and methodology induced the motivation to have this book. Regardless the endless representation styles and techniques, we would like to propose a simple way: returning to your own ideas, and returning to the essence of the space, it may be a more profound choice. The interesting thing is that when discussing with Valerio about his teaching, we discovered the similarity between his "extended representation" and Chinese traditional paintings, namely, the poetry. Maybe you have already noticed. So never take it for granted or easily abandon the treasure you already have.

Shuping Xiang
Francesco Belligerante

**图书在版编目(CIP)数据**

景观的延伸表现——瓦莱里奥·莫拉比托的理念与方法/项淑萍，（意）弗朗切斯科·贝利杰兰特. —北京：中国建筑工业出版社，2013.10
ISBN 978-7-112-15817-1

I. ①景… II. ①项…②贝… III. ①景观设计－教材
IV. ①TU986.2

中国版本图书馆CIP数据核字（2013）第209584号

责任编辑：徐　纺　滕云飞

**景观的延伸表现——瓦莱里奥·莫拉比托的理念与方法**
项淑萍　[意]弗朗切斯科·贝利杰兰特
*
中国建筑工业出版社出版、发行（北京西郊百万庄）
各地新华书店、建筑书店经销
江苏恒华传媒有限公司制版
北京缤索印刷有限公司印刷
*
开本：889×1194毫米　1/20　印张：10½　字数：252千字
2013年10月第一版　2015年4月第二次印刷
定价：68.00元
ISBN 978-7-112-15817-1
（24579）